Lean Supply Management

Ade Asefeso MCIPS MBA

Copyright 2015 by Ade Asefeso MCIPS MBA
All rights reserved.

First Edition

ISBN-13: 978-1507609156

ISBN-10: 1507609159

Publisher: AA Global Sourcing Ltd
Website: http://www.aaglobalsourcing.com

Table of Contents

Disclaimer 5
Dedication 6
Chapter 1: Introduction 7
Chapter 2: Lean Green Manufacturing 15
Chapter 3: Vendor Survey and Qualification 19
Chapter 4: Let Lean Guide You to Green 23
Chapter 5: Extending Lean and Green to the Supply Chain 25
Chapter 6: Lean and Green Enterprise 29
Chapter 7: Five Reasons Why Sustainability and Supply Chain Greening Will Stick. 37
Chapter 8: Green Branded Shipping Options for E-Commerce 41
Chapter 9: Solving the Sustainable Sourcing and Green Supply Chain Management Puzzle 45
Chapter 10: Greener Shipping and Logistics 49
Chapter 11: A Sustainable Transport Plan 53
Chapter 12: Green Supply Chain Brings Competitive Advantages 55
Chapter 13: Sustainable Supply Chains 61
Chapter 14: Motivating Suppliers to Meet Sustainable Sourcing Requirements 65
Chapter 15: Sustainable Procurement 69
Chapter 16: Game On to Drive Sustainable Procurement 75

Chapter 17: Future-Proofing Supply Chains With Sustainable Procurement 81
Chapter 18: Sustainable Packaging 85
Chapter 19: Designing Sustainable Products and Services 93
Chapter 20: Developing New Business Models 97
Chapter 21: Sustainable Supply Chain Alignment 101
Chapter 22: Watch Your Step 105
Chapter 23: Viewing Compliance as Opportunity 109
Chapter 24: Sustainability Return on Investment 115
Chapter 25: Making Value Chains Sustainable 117
Chapter 26: Creating Next Practice Platforms 123
Chapter 27: Navigating Sustainable Supply Chain Management in China 127
Chapter 28: Sustainable Supply Chain Creates a Competitive Advantage Worldwide 135
Chapter 29: German Sustainable Development Strategy 141
Chapter 30: Developing a National Sustainable Good Greenhouse Gas Inventory 143
Chapter 31: Hope for Sustainable Economies 147
Chapter 32: Green Supply Chain Management Requires Less Procrastination 153
Chapter 33: Triple Bottom Line 157
Chapter 34: Conclusion 169

4

Disclaimer

This publication is designed to provide competent and reliable information regarding the subject matter covered. However, it is sold with the understanding that the author and publisher are not engaged in rendering professional advice. The authors and publishers specifically disclaim any liability that is incurred from the use or application of contents of this book.

If you purchased this book without a cover you should be aware that this book may have been stolen property and reported as "unsold and destroyed" to the publisher. In this case neither the author nor the publisher has received any payment for this "stripped book."

Dedication

To my family and friends who seems to have been sent here to teach me something about who I am supposed to be. They have nurtured me, challenged me, and even opposed me.... But at every juncture has taught me!

This book is dedicated to my lovely boys, Thomas, Michael and Karl. Teaching them to manage their finance will give them the lives they deserve. They have taught me more about life, presence, and energy management than anything I have done in my life.

Chapter 1: Introduction

The elimination of waste everywhere while adding value for customers. This definition is a natural fit with sustainability and the "Lean and Green" business ethic. Lean manufacturing has demonstrated how companies have saved or avoided enormous operating and maintenance costs and significantly improved the quality of their products.

Lean manufacturing looks at manufacturing from a systems perspective, which includes a thorough evaluation of upstream and downstream process inputs and outputs. Viewed this way, suppliers and customers play a critical role in successful lean manufacturing. Heavy emphasis is placed on design and innovation and obtaining input from supply chain partners, individuals and organizations through a process called 'value-stream mapping'.

The Lean, Green and Supply Chain Intersect

Lean efforts have been demonstrated to yield substantial environmental benefits (pollution prevention, waste reduction and reuse opportunities etc.) however, because environmental wastes and pollution are not the primary focal points, these gains may not be maximized in the normal course of a lean initiative. This is because lean waste is by its nature not always in sync with typical environmental wastes. I argue that by looking deep into your value chain (upstream suppliers, operations and end of life product opportunities) with a 'green' or environmental lens, manufacturers can eliminate even more waste in the manufacturing process, and realize some potentially dramatic savings

Where 'lean' creates a positive view (future state) of a process without waste, 'green' creates an alternative view of a sustainable future for organizations that play in the global marketplace or offer a unique disruptive innovation. Lean and green approaches to manufacturing not only leverages compliance issues but also puts companies on the path to going beyond compliance.

To make progress on environmental issues, organizations must understand that they are part of a larger system. To confront these issues practically, you need employees who are innovative; who have the skill and the vision to redesign products, processes, business models and who understand the business context. Most important, they need to be able to tell a story about why this is a meaningful journey.

If they are stuck in the mind-set (so popular in business schools, unfortunately) that a company exists to maximize return on investment capital, with an emphasis on short-term financial performance, they won't get very far.

To me; it starts with "Who are we?" and "Why are we here?" In a great book called The Living Company, published in 1997, Arie de Geus described a study conducted by Shell in the early 1980s of companies that had survived for more than 200 years. What those organizations had in common was an understanding of themselves as a human community first and a machine for making money second.

We have lost a lot of that sense of company as community. But a business is a group of people working together. You can't build a new vision without a strong, community-oriented culture which, by the way, is rare. It sounds good, everybody nods their heads, but the gap between that idea and the way most managers manage is enormous.

The challenges that must be overcome to make a business more holistic from end to end.

The first challenge is to understand the larger system you are in. The second is to learn to work with people you haven't worked with before. Those two skills might seem distinct, but in practice they are interwoven. The system is too complicated for one person to grasp. It crosses too many boundaries, both internal and external.

The third challenge has to do with how you perceive sustainability. They might not say this, but most companies act as if sustainability is about being less bad. There is certainly a need to reduce your carbon footprint. But people don't get excited about incremental changes like that. They need a more ambitious vision.

The environmental movement is a big culprit in this. There has been so much rhetoric about how bad business is that people inside companies feel guilty, and guilty people aren't going to do bold things. But that is crazy! Innovation is what good businesses do best. They are all about creating new sources of value. NGOs and governments can't possibly solve these problems alone; business innovation is essential.

Making change across a supply chain

First, you focus on the nature of the relationships. In most supply chains, 90% of them are still transactional. If I am a big manufacturer or retailer, I pressure my upstream suppliers to get their costs down. There is very little trust and very little ability to innovate together. That must change, and it is starting to.

Second, you learn to work with NGOs and other non-business entities. They will give you access to expertise that you can't grow fast internally. Water is a classic example. A few years ago, Coca-Cola decided to cut the water used to make a litre of Coke from more than three litres to 2.5 litres. But it was overlooking the 200-plus litres it took to grow the sugar that went into that Coke. The company found that out because it partnered with the World Wildlife Fund, which knew how to analyze the water footprint of the value chain. Coca-Cola now knows the difference between drip-irrigated sugarcane and flood-irrigated sugarcane.

Corporations need NGOs.

For credibility, especially in Europe. People don't trust the business-as-usual mind-set for good reason.

I don't think it's so different in the U.S. If a credible NGO certifies your product, your brand can gain hugely if you are willing to change your practices. NGOs can also provide knowledge. No business knows what Oxfam knows about the plight of farmers or what WWF knows about biodiversity and watersheds. The best businesses don't just hire the sharpest people; they also keep expanding their expertise by partnering with NGOs that have deeper and broader knowledge.

Roles of leaders in confronting supply chain challenges.

These are all leadership issues. But when I say "leader," I don't necessarily mean the CEO or even bosses in general. You can't possibly source everything sustainably, as Unilever has declared as a 2020 goal, unless you engage thousands and thousands of people around the world. You will need technical innovations, management innovations, process innovations, and cultural innovations. The people who figure those out are leaders, by definition, and most of them won't be senior executives. All the word "lead" means, if you look at its Latin root, is to step across a threshold.

Does leadership ever come from outside the organization?

Sure. When someone comes into an organization; a new hire or even a new supplier; he or she will ask, "Why do we do it this way?" The answer is often "Just because." Now, 90% of those habits may be perfectly okay. But 10% are completely dysfunctional, particularly when the world around you is changing. A cool part of sustainability work is uncovering the assumptions that lead people to do things in a way that's out of touch with the company's larger reality.

Getting the ball rolling inside organizations.

In some cases, it starts when the CEO steps back to reconsider the organization's relationship to the world. Or when people deep in the business discover some problems, collect some data, and then try to find people who have the skills to do something. Gradually, they build up a network of people who are fascinated by the problem and excited about finding a solution. Nothing motivates designers more than being told that something's impossible, like eliminating a toxin that has always been part of a product or process. That started to happen after Nike's first toxicological study of its products and supply chain more than 10 years ago.

The key is not your position; it's your passion, your ability to form networks, and your organizational savvy. This is where young people sometimes run aground; they have got energy and passion but don't have a clue about the culture of the organization or where the natural pockets of power are.

Obviously, if you are talking about operational changes, you need operating managers. Things get a lot more real when you talk about negotiating and writing contracts differently and changing metrics to gauge more-sustainable sources.

For more than a century, financial historians have used the "bubble" metaphor to explain why smart people act really stupidly together. Inside the bubble, there is a reality, a language. But outside the bubble there is a larger reality that asserts itself eventually and the bubble bursts. If you think of the industrial age as a bubble, then maybe the larger realities (like finite resources) are starting to assert themselves.

Some people think the industrial age is already over; we are now in the information age. But that is a serious misunderstanding. The industrial age has always been punctuated by radical shifts in dominant technologies; the electric light, the automobile. The internet is just the most recent. The industrial age has been the era when machines and machine thinking shaped our lives. Leaders need to imagine life and their businesses outside that bubble, where efficiency, productivity, and the maximization of return on capital are balanced by the imagination, passion, and trust that shape creativity and innovation.

Chapter 2: Lean Green Manufacturing

If you have ever sat in traffic having simultaneous thoughts that "all of these vehicle emissions are bad for the environment," and "I wish I lived closer to work," then you already have a sense of how mutual "lean and green" thinking works.

The traditional way of thinking goes that "green" business initiatives add costs, while implementing "lean" processes is about streamlining and saving money.

Many manufacturers today have evolved their thinking so that lean and green initiatives work hand in hand, achieving the same goal of increasing profits.

In fact, recent study suggested there is a "synergistic relationship" between lean and green systems, and that there are "philosophical and structural similarities" between the two models.

Lean manufacturers follow stringent manufacturing processes designed to eliminate or minimize waste and non-value added steps in seven categories.

You can think of them as the "seven deadly sins" of wasteful manufacturing: Defects, Overproduction, Transportation, Waiting, Excess Inventory, Unnecessary Movement and Over-Processing. Some manufacturers have also adopted a "Six Sigma" improvement process, which includes a set of disciplined tools and problem-solving methodologies for reducing or eliminating process variation and product defects.

Here are seven examples of how these lean innovations can yield sustainability results for manufacturers.

1. Fewer product defects: If you have improved your processes to minimize product defects, that means you are using fewer raw materials to manufacture those products. In addition, you don't need as much plant space, systems and equipment to rework or repair those products, which equals less energy consumption.

2. Less overproduction: Overproduction means manufacturing in excess of your customer orders. Eliminating overproduction is a major focus of lean. In traditional manufacturing reasoning, if a production line is running and you've already made all of the products to meet customer demand, you make more of something to justify the expense of your equipment and people. Lean concepts require that you only produce what you need, when you need it. If you don't overproduce then you consume fewer raw materials, use less energy to operate, and eliminate the risk associated with not selling the excess inventory and eventually disposing of it as waste.

3. Minimizing wasted movement: A great example of a wasteful motion is when a production area is poorly designed so that workers are wasting time and effort lifting things unnecessarily or needing to walk an excessive distance back and forth to find tools or complete a task. An ineffective layout requires more space increasing heating, cooling, and lighting demands. It can also increase the time to produce a product resulting in increased energy requirements.

4. Reducing transportation: An example of wasted transportation is by having your production facilities not located near your customers, requiring that you transport materials over long distances. It can also relate to the movement of materials within your facility. Internal movement of materials adds no real value to the product, but increases the energy used and the costs associated with the product. Lean thinkers look to minimize transportation wherever possible.

5. Less excess inventory: Similar to overproduction, if you have less product inventory sitting around, you can use your plant space more efficiently (saving heating and cooling demands) while also consuming less packaging and raw materials. Lower levels of inventory also reduce the risk of waste due to obsolescence and undiscovered defects.

6. Reduced waiting: Nobody likes waiting, especially those of us who are lean thinkers. A key lean concept is reducing waiting for things like equipment to be available, information, or materials. A great example of waiting is when your production processes aren't balanced, so when an operator has finished part of a task, he needs to wait for a machine to complete a cycle before finishing that task. Syncing up these processes to reduce waiting can cut down on production downtime, which means you have less wasted energy.

7. Less over-processing: Over-processing means you are adding more steps or materials to something than what the customer will pay for. In other words, every step of a production process should add customer value. Improving your processing to just what is needed allows you to cut down on waste and lower your environmental footprint.

Applying lean thinking to your sustainability efforts will help ensure that your green initiatives will have long-term staying power because of the added value to your business.

If you are a manufacturer seeking to do the right thing for the environment for altruistic reasons, you may be able to achieve this goal more effectively by implementing lean systems. Green initiatives are going to be more justifiable long term if they create more success for your business. Establishing a culture that embraces both concepts can attain significant results.

Chapter 3: Vendor Survey and Qualification

Manufacturers also supplement their Lean efforts by surveying their supply chain partners and asking a series of questions designed to identify where the resource consumption and waste management opportunities may lie. These questions will help determine if technology, operational practices, enhanced training and awareness or other tools can make their company more sustainable and lead them down the path to make the decision that best meets their business needs. These questions include but are not limited to.

1. How can I leverage my manufacturing capabilities and processes in a way that optimizes per unit material resource consumption?
2. Can I reduce waste generation through improving material use, scrap/off spec reuse and improved equipment maintenance?
3. Can I work collaboratively with my intermediate parts or materials suppliers to use life cycle design practices and manufacture parts with lowered environmental footprints?
4. How can I encourage suppliers to increase equipment efficiency, reduce manufacturing cycle time, reduce inventories, streamline processes or seek quick returns on investment?

5. Can I improve my sales and operations planning to optimize production runs and reduce resource loads or generated wastes?
6. How can I work more closely with logistics and transportation partners to optimize shipment schedules, customer deliveries, warehousing, routing and order fulfilment?
7. Can I work with my customers and product designers to improve packaging to optimize space reduce materials use and improve load management?
8. How can I collaborate more closely with customers to enable reverse logistics and profitable product reusability?
9. What types of value-added training and development programs can I develop to promote lean and green opportunities with my suppliers?

Lean-Green Synergies Are Not Without Challenges

Sometimes there are potential conflicts with certain types of lean strategies leading to changes in supply change management. For instance, lean strategies that require just-in-time delivery of small lot sizes require increased transportation, packaging, and handling that may contradict a green approach. Introducing global supply chain management into the green and lean equation increases the potential conflict between the green and lean initiatives.

So as companies begin to implement lean and green strategies in supply chains, especially large and complex global supply chains, manufacturers need to explore the overlaps and synergies between quality-based lean and environmentally based 'green' initiatives, and understand the various trade-offs required to balance possible points of conflict. If your organization have been reluctant to engage your supply chain or implement or maintain environmental initiatives in your product manufacturing because of the perception that you can't afford it, then think again. It is more likely that you cannot afford to ignore it.

Chapter 4: Let Lean Guide You to Green

I meet with too many organizations that are still addressing sustainability as if it's something new, something different, when in fact many of them have been working on it for over 20 years. They just know it by a different name, Lean.

Many of those companies are surprised to learn that they can take the tools, systems thinking and lessons learned from the process improvement methodology, Lean, and apply them effectively to the operationalization of sustainability.

What happens when a company adopts Lean? Processes are studied, problem-solving teams are established, process improvement methodologies are adopted, and employees accept new responsibilities and boundaries to improve their organizations and connect to customers and other stakeholders outside their traditional work responsibilities. Goals, such as "Zero Defects," are adopted with tenacity.

We are lucky enough to live in a time of a new Industrial Revolution. Sustainability is challenging us to re-examine everything, including the former linear model of Take-Make-Use-Waste and, instead, explore a circuitous model of Borrow-Make-Use-Return that will, theoretically, have no waste. Waste, by-products that are not inputs to another process, will then be viewed as a defect.

When companies expand the definition of waste to include not only product and process waste, but also the business consequences of unsustainable practices, Muda's list of wastes takes a different form.
1. Waste of natural resources.
2. Waste of human potential.
3. Waste due to emissions.
4. Waste from by-products (reuse potential).
5. Terminal waste, that is by-products that have not further usefulness.
6. Energy waste.
7. Waste of the unneeded (e.g., packaging)

When sustainability is viewed this way, it isn't something new that has to be planned from scratch and agonized over. Instead, it can be integrated into existing continuous improvement programs; Lean, or even Six Sigma initiatives.

When the definition of waste is expanded and when it's understood that the consequences of corporate decisions extend past the company parking lot, Lean can indeed be green.

Chapter 5: Extending Lean and Green to the Supply Chain

Becoming a green organization as part of a lean initiative occurs sometimes by design, and sometimes by accident. A recent research found some interesting results when evaluating how lean manufacturing, sustainability and supply chain management may at times be complementary. The study found, among other things that:
1. Firms tend to have more sophisticated lean strategies than green strategies, and because of this awareness of 'sustainability' in supply chain management circles is less mature.
2. Lean and green initiatives overlap, where projects that meet lean objectives often provide unanticipated green benefits.

Establishing initial goals for manufacturing efficiencies include maximizing parts, machine and material utilization, human movement and of course reducing waste. This series of continuous improvement steps offer a cornerstone for reaching both a green and efficient supply chain. But how can manufacturers work beyond the 'four walls' of their organizations to green their supply chain? A green focus in supply chain management requires working with upstream suppliers and downstream customers, performing analyses of internal operations and processes, reviewing environmental considerations in the product development process, and looking at extended stewardship opportunities across the lifecycle of one or more intermediate or final products.

Lean Tools You Can Use

So far, I have laid a foundation for Lean Manufacturing and the intersection with supply chain management. This next section presents a couple of widely accepted practices that are used in Lean design and manufacturing, which can be modified to capture supplier network considerations.

1. Value-Stream Mapping

A strategic approach to mapping environmental and lean opportunities would be to map the 'value-stream' of one or more products as a way to seek where the greatest waste reduction and environmental impact reduction opportunities are. Value stream mapping arrived on the business process landscape with the emergence of Lean engineering, design and manufacturing. A process and systems based methodology, value stream mapping can help organizations to identify major sources of non-value added time and materials resources i.e. waste that flow into the manufacturing of a particular product or (even) service; and to develop an action (or "Kaizen") plan to implement less wasteful practices and processes. From an environmental perspective, practitioners can also look at processes from an environmental, health and safety point of view, focusing on processes tending to use great amounts of resource inputs and that generate waste outputs.

Typical steps in value stream mapping include:

1. Select a product or process(es).
2. Through interviews and work observations, collect data on the 'current state' of the value stream (inputs and outputs).
3. Using a cross functional team (CFT) of knowledgeable staff, develop a 'current state' value stream map; focus on identifying over consumptive or waste generating activities.
4. With the CFT in place, brainstorm ideas to improve resource use, production flow, waste capture and reduction, reuse and off spec material reuse, and labour/time management.
5. Create a future state' value stream map that identifies areas, targets and key performance metrics for continual improvement.
6. Develop a implementation plan, complete with authorizes and responsibilities.
7. Develop continual improvement measurement and monitoring program.
8. Last but not least...get started!

A company that works more sustainably is executing a good business practice because it forces the company to pay more attention to wasting less and conserving more, which increases profits in the end. When you conserve things, when you don't waste things, when you are frugal with the use of resources, actually you find that you can produce more for less cost, and that is more profitable. It's good business sense. For Example a major UK retailer adopted a sustainable business practice by setting goals that included cutting its carbon omissions in half by 2020 and cutting them completely by 2050, two goals that the company is on target to meet. To do that, they had to start with-in stores because that is where most of the company's carbon omissions came from. So they outfitted their stores with new lighting and refrigeration systems, new glass and different power systems, and it started building stores out of wood. The company also had to get its manufacturers, farmers and other pieces of the supply chain on board with the plan, and set a goal of reducing their carbon omissions by 30% by 2020.

Including sustainability in companies' lean thinking practices also makes them more relatable to their employees. They don't like to waste things, throw things away. So they appreciate it when they see the organization trying to do the right thing.

Chapter 6: Lean and Green Enterprise

I thought that in light of the economic body slamming that has been going in the past years, it's worth reflecting on some efficiency-based ways that businesses can use to overcome (or at least buffer) some of the external factors that are causing such economic uncertainty. Like the hikers seeking shelter from the storm, there are some "lean-to" like steps that company's can take to exert some control and influence and it all relates to a leaner, greener, smarter enterprise.

In one of my books "Green Manufacturing (Paradigm Shift to Sustainable Capitalism)" I wrote about how importance a "lean and green" enterprise was in establishing a smarter, leadership position in a rapidly changing global marketplace. I noted then that a study suggested that "lean companies are embracing green objectives and transcending to green manufacturing as a natural extension of their culture of continuous waste reduction, integral to world class Lean programs." Lean was more rapidly accomplished with a dedicated corporate commitment to continual improvement, and incorporating 'triple top line' strategies to account for environmental, social and financial capital. I also argued by looking deep into an organizations value chain (upstream suppliers, operations and end of life product opportunities) with a 'green' or environmental lens, manufacturers can eliminate even more waste in the manufacturing process, and realize some potentially dramatic savings

So I was reminded this past week that Lean in design, Lean in manufacturing, and Lean in inventory can individually or collectively be key success factors in managing waste in all its many forms. Collectively, this can have a measurably positive effect on a company's financial, and hence, business performance.

Lean Design

I came across an older but very relevant article written in the aftermath of the Internet stock crash in the early 2000's. The article described product development as involving "two kinds of waste; that associated with the process of creating a new design (e.g., wasted time, resources, development money), and waste that is embodied in the design itself (e.g., excessive complexity, poor manufacturing process compatibility, many unique and custom parts)." The article cautioned that because the design process is the cradle of creative thinking, designers needed to carefully watch what they "lean out" or risk cutting off the creative process to reduce waste. What has happened in the ensuing years has been an incredible emphasis on "green design" that focuses on full product life cycle value, such that "end of life management" considerations have taken on a more relevant and embedded nature in manufacturing.

A Lean Manufacturer Can be a Sustainable Manufacturer.

As stated earlier in this book; Lean manufacturing practices and sustainability are conceptually similar in that both seek to maximize organizational efficiency. Where they differ is in where the boundaries are drawn, and in how waste is defined. As I have in one of my books "5s Lean Manufacturing (Key to Improving Net Profit)", that Lean manufacturing practices, which are at the very core of sustainability, save time and money an absolutely necessity in today's competitive global marketplace. The key areas to control manufacturing waste and resource use during the design and manufacturing cycle, can be broken down and managed for waste management and efficiency in the following five ways.

1. Reduce Direct Material Cost: Can be achieved by use of common parts, common raw materials, parts-count reduction, design simplification, reduction of scrap and quality defects, elimination of batch processes, etc.

2. Reduce Direct Labour Cost: Can be accomplished through design simplification, design for lean manufacture and assembly, parts count reduction, matching product tolerances to process capabilities, standardizing processes, etc.

3. Reduce Operational Overhead: Efficiencies can be captured by minimizing impact on factory layout, capture cross-product-line synergies (e.g. a modular design/mass-customization strategy), improve utilization of shared capital equipment, etc.

4. Minimize Non-Recurring Design Cost: Planners and practitioners should focus on platform design strategies to achieve efficiencies, including; parts standardization, lean QFD/voice-of-the-customer, Six-Sigma Methods, Design of Experiment, Value Engineering, Production Preparation (3P) Process, etc.

5. Minimize Product-Specific Capital Investment through: Production Preparation (3P) Process, matching product tolerances to process capabilities, Value Engineering / design simplification, design for one-piece flow, standardization of parts.

Lean Inventory Management can Drive Sustainable Resource Consumption.

It is estimated that the global marketplace is sitting on $8 trillion worth of 'for sale' inventory. The U.S. maintains a quarter of that inventory. These idle goods not only represent a tremendous financial burden but an enormous environmental footprint that was generated in the manufacturing of those goods. If we could permanently reduce the amount of product sitting idle, we did save money, energy, and material. The problem is predicting and managing inventory in such fickle times. New predictive tools being advanced by companies that hold promise in nimbly driving inventory demand response up the supply chain. For instance, using both demand sensing software and good management practices, P&G has cut 17 days and $2.1 billion out of inventory. All that production avoided saves a lot of money in manufacturing, distribution, and ongoing warehousing. It also saves a lot of carbon, material, and water.

What we found shocking was that even with the fastest-selling, most predictable products, the estimates are off by an average of more than 40 percent. Imagine that a CPG company believes that 2 million bottles of a fast-turning laundry detergent will sell this week. With 40 percent average error, half the time sales will actually fall between 1.2 million and 2.4 million bottles and the other half of the time sales will be even further off the mark. The process becomes self perpetuating and the inventory racks up along with the parallel environmental footprint, unless somehow the uncertainty can be better predicted. While companies like to have on hand what we referred to as "safety stock", I have come to know as reserve inventory driven by "just in time" ordering. But that process was shown to have its own flaws such as when orders for goods dried up overnight in 2008 and when it came time to ramp up in early 2010, part counts were insufficient to meet the rising demand.

I really pity the supply chain demand planner, who like the weatherman is subject to the fickle nature of an unpredictable force. Reducing the inventory itself could be the greenest thing logistics executives can do. I had the chance to speak and attend a recent Supply Chain Summit where demand response planning was discussed at length and where green supply chain issues were recognized as one of many key attributes in effective supply chain management. In such a volatile economy, its vital that companies keep inventory management in mind as a way to leverage its costs and simultaneously look toward environmental improvements that can reduce waste.

Partnering for Progress

A relatively recent pilot program in the State of Wisconsin just shows how partnering to create a lean focused sustainable manufacturing cluster can have enormous dividends. According to a recent article in BizTimes.com, the Wisconsin Profitable Sustainability Initiative (PSI) was launched in April 2010 by the Wisconsin Department of Commerce and the Wisconsin Manufacturing Extension Partnership (WMEP). The goal according to the article is "to help small and midsize manufacturers reduce costs, gain competitive advantage and minimize environmental impacts". Forty-five manufacturers participated in over 87 projects evaluated. These projects focused on "evaluating and implementing a wide range of improvements, including reducing raw materials, solid waste and freight miles, optimizing processes, installing new equipment and launching new products. The initial results show that the projects with the largest impact do not come from the traditional sustainability areas such as energy or recycling. Instead, outcomes from the initial projects suggest that transportation and operational improvements are places where manufacturers can look to find big savings, quick paybacks and significant environmental benefits.

The program is projected to generate a five-year $54 million economic impact, including: $26.9 million in savings, $23.5 million in increased/retained sales and $3.6 million in investment.

Together, lean design, lean manufacturing and effective, lean inventory management offer a "trifecta" approach for industry to identify, reduce or eliminate and track waste. Effective use of these tools cannot only drive both in how the product is designed and produced but offers opportunities all the way up the supply chain to manage effective inventory and resource consumption.

Chapter 7: Five Reasons Why Sustainability and Supply Chain Greening Will Stick.

1. Economics: Contrary to popular belief, making the business case for making sustainability operational within an organizational supply chain is becoming easier, not harder. With the availability of more data from 'first movers', procurement managers, environmental directors, design engineers, marketing/communications staff and operations managers (among others) are now able to make strong business cases in favour of looking at operations through a green lens. In addition, barriers to global trade brought on by increasing environmental regulations, more stringent restrictions on hazardous substances, greater emphasis on lean manufacturing, and increased supplier auditing and verification are creating the critical mass toward a new norm in supply chain management and expectations. Seeking efficiencies in supply chain management and producing products while reducing waste continue to be a vital imperative in a recovering economy. Those who neglect to critical evaluate their operations from a sustainability point of view this year will be cast to the side.

2. Climate Action: Supply chain sustainability is affecting shareholder value, company valuations and even due diligence during proposed mergers and acquisitions. Shareholder actions on sustainability performance and transparency were up 30% in 2014.

A recent insightful report note that "As carbon pricing becomes established in various jurisdictions, organizations will face risks from compliance obligations. This will impact cash management and liquidity, and carbon-intensive sectors may see an increase in the cost of capital." Still much work still remains to infuse green thinking in the C-Suite. Little more than a third of those executives surveyed indicated that they were working directly with suppliers to reduce their carbon footprint, or have just started discussing climate change initiatives with their suppliers and now, the World Resources Institute is completing authoritative new supply chain and product lifecycle greenhouse gas protocols that will frame what is expected to be a burgeoning wave of value chain sustainability accounting and reporting. Stay tuned!

3. Disclosure and Accountability: As I have previously noted, supply chain management became widely recognized in 2014 as a key factor in measuring the true "sustainability" of an organizations practices and processes, and ultimately its product or service. Increased attention will be paid this year on conflict minerals (because of the passage of the Dodd-Frank Wall Street Reform and Consumer Protection Act of 2010), fair labour and other social aspects of sustainability, ongoing management of hazardous substances in toys and other consumer products, and looking at the supply chain to manage risks and liabilities from product recalls and other environmental impacts of products and services. The concept of "materiality" in corporate social responsibility and product disclosure (FTC Green

Guidelines) and SEC financial reporting is taking on new meaning from a supply chain perspective. 'Materiality' in terms of supply chain or network management will require more rigorous implementation and oversight of ethical business practices and practicing proactive environmental stewardship through-out a products value chain. Suppliers play a key external role in managing the environmental, social or financial issues within the product value chain.

4. Innovation and Collaboration: The emergence of collaborative opportunities among larger manufacturers creates entry points in the market for smaller, intermediate products manufacturers as well. Larger companies are identifying the critical supply chain partners that have the greatest product impact and begin seeking ways to collaboratively address the environmental and social footprint of their products through the value chain. A new report even suggests that consumers will play a leading role behind greater supply chain collaboration. The report suggests that while suppliers are independently seeking more open, collaborative ways to move goods, consumers may be "the trigger for an optimized collaborative supply chain flow; this next level of supply chain optimization is based on transparency and collaboration." More specifically, "Consumer awareness about sustainability demands a more CO_2-friendly supply of products and services".

5. Life Cycle Design and End-of-Life Product Management: There are increased challenges that the waste management industry is facing, wider attention paid to greener packaging and increased emphasis on financial accountability is being felt in world markets. Establishing a reverse logistics network that supports life cycle design, Extended Producer Responsibility (EPR), and "demanufacturing" processes will take on higher meaning in 2015. According to a recent white paper issued by a sustainability expert; EPR is a market-based approach that effectively assigns end-of-life responsibility and product stewardship to producers, requiring them to meet specific targets for material recycling and recovery, relative to the total amount of packaging that they have put into the marketplace.

EPR helps to shift the responsibility for collecting packaging and end of life products from financially tapped out local government to producers. But upstream of the manufacturing process, EPR success can be achieved through incentives for companies to take a closer look at how they design products for better end-of-life management (life cycle design). Producers are not alone in addressing the social and ecological impacts of their products. Manufacturers must engage their supply networks to help drive EPR upstream; however, downstream customers play a role too. So producers and consumers should strive in 2015 to continue a dialogue about what to do to improve the profile of consumer products in a way that is a win-win for all affected stakeholders.

Chapter 8: Green Branded Shipping Options for E-Commerce

There are still e-commerce and subscription box shippers who believe that they have to choose between sustainability and good looking, eye catching branding. Fortunately that is not true and there are several options available to you that will deliver your product to your customer in a package that communicates your brand and message, especially if it happens to be a green one.

Here are a few options that you may want to consider.

1. Paper mailer envelopes: Even though I personally am not a great fan of padded mailers because they tend to be heavy and sometimes messy, unpadded mailers can be made of 100% recycled content, and are 100% recyclable. They can be easily recycled at curb side or with any other paper waste.

For many products such as soft goods and apparel that do not require the protection of a box, paper mailer envelopes are a great option that can be printed in two colours in minimal quantities and at minimal cost.

2. Plastic mailer envelopes: Even though they may not be perceived as a green packaging solution by some consumers, they add very little weight and of course you can't beat the moisture protection they provide. Plastic mailers are available in a variety of film structures, even some with a high percentage of recycled waste content and are typically 100% recyclable wherever plastic bags are recycled.

Even though plastic mailer envelopes usually require much higher minimums for printing, the graphics can be sharper with more colours and finishes available compared to flexographic printing on paper.

3. Corrugated shipping containers: Provide the most options with many designs available including RSC style shipping boxes, FOL end/side load mailers, and even paper tubes. The wide variety of die cut mailer boxes includes top tuck, front lock and application permitting, Eco-Mailer boxes that typically use about 30% less board compared to other designs.

The available recycled content varies depending on the type and style being used but most have at least 30% recycled content, and some as much as 100% PCW content. Most corrugated containers can easily be recycled with other paper waste.

Though most corrugated containers used for e-commerce are one or two colours, screens and reverse printing designs make it easy to create a low cost and unique graphic design. Also, interior printing has become very popular, increasing the available billboard and message/logo placement where it is sure not to be missed.

Which type of shipping container is best for you is determined by a number of factors including what you are shipping and how you are shipping it, as well as the level of protection your product requires. An increasingly important factor is shipping weight and how it impacts shipping costs. A client of ours reduce customer's shipping cost by two dollars per box by analyzing their box design, the board composition and even ink content. Their carbon footprint was reduced, the cost per box was also lowered, and the shipping cost savings was over six figures, per month.

In many cases however, the deciding factor may be what type of impression you want to make on your customer in terms of style, graphics, and how green-minded that customer is. Would we suggest sending organic cotton baby clothing in a plastic mailer? Probably not.

Every application is a little different so there are no "one size fits all" solution however, there is a low cost, attractive and eco-friendly packaging solution for every e-commerce company.

Chapter 9: Solving the Sustainable Sourcing and Green Supply Chain Management Puzzle

2014 has been truly remarkable in a number of key areas of green supply chain management from a number of perspectives, including: policy and governance, operations and optimization, guidance and standardization and metrics. The green pieces of the supply chain and sustainability puzzle appear to be nicely falling into place. Key themes that I can glean from this most incredible year are.

1. Big Industry Movers and Government Green up the Supply Chain

Over the past year; observers and practitioners read nearly weekly announcements of yet another major manufacturer or retailer setting the bar for greener supply chain management. With a much greater focus on monitoring, measurement and verification, Wal-Mart, IBM, Proctor and Gamble, Kaiser Permanente, Puma, Ford, Intel, Pepsi, Kimberly-Clark, Unilever, Johnson & Johnson among many others made a big splash by announcing serious efforts to engage, collaborate and track supplier/vendor sustainability efforts. Central to each of these organizations is how vendors impact the large companies carbon footprint, in addition to other major value chain concerns such as material and water resource use, and waste management. Even government agencies in the U.S. (General Services Administration) and (DEFRA in Britain) have set green standards and guidelines for federal procurement. More and more companies are jumping on the green train and the recognition is flowing wide and deep.

2. Supply Chain Meets Corporate Social Responsibility

Adding to many companies existing concerns over environmental protection, large products manufacturers such as Nestle, Corporate Express, Danisco, Starbucks, Unilever and the apparel industry stepped up in a big way to address human rights, fair labour and sustainable development in areas in which they operate throughout the world. Each of these companies and others like Wal-Mart have embraced the "whole systems" approach; that underscore transparency and collaboration in the supply chain. Each company recognizes that to be a truly sustainable organization, it must reach deep beyond its four walls to its suppliers and customers.

3. Transparency and Collaboration Take on a Green Hue

Suppliers and customers can collaboratively strengthen each other's performance and distributing cost of ownership. Practitioners have found "reciprocal value" through enhanced product differentiation, reputation management and customer loyalty and the continuing Wiki leaks controversy is boldly reminding the business world that accountability and transparency and corporate social responsibility is vital and may even be a game changer in how products and services are made and delivered to the global marketplace.

4. Logistics Turning to Greener Solutions

Numerous studies and surveys conducted by peer organizations this year underscored how sustainability among carriers and shippers was central in the minds of most logistics CEO's. Whether it was by land, air or sea, shipping and logistics embraced sustainability as a key element of business planning and strategy in 2014. I also had the pleasure of visiting FedEx's and learned of the myriad of operational innovations and sustainability focused metrics that the company is tracking throughout its operations and maintenance activities. UPS even mentioned its efforts to manage its carbon footprint in its catchy new brand campaign "I Love Logistics". Finally logistics companies are partnering with manufacturing to support reverse logistics efforts designed to manage end of life or post consumer uses of products or resources.

5. Lean Manufacturing Meets Green Supply Chain

As manufacturing continues its slow rebound from the Great Recession, companies are recommitting themselves to implementing less wasteful production as a way to leverage cost and enhance savings. Parallel efforts are in play also to incorporate more environmentally sustainable work practices and processes. Enhancing this effort to lean the product value chain is recognition of upstream suppliers and vendors work practices and possible impacts they may have on manufacturing outputs. Lean efforts have been demonstrated to yield substantial environmental benefits (pollution prevention, waste reduction and reuse opportunities) as well as leverage compliance issues. More and more, companies are exploring the overlaps and synergies between quality-based lean and environmentally based 'green' initiatives.

6. Looking Forward to 2015

Yes indeed, it's been a big year for supply chain management and its intersection with sustainability. I see little for 2015 that will slow down this upward green trajectory, and naturally I am glad. I am glad that more businesses "get it" and don't want to be viewed as laggards in leaning towards a business ethic that values sustainability and socially influenced governance. I am glad that more companies are seeking out green innovation through new technologies and being 'first movers' in their respective business spaces and I am glad that you (my readers) and I am here to be part of the change.

Chapter 10: Greener Shipping and Logistics

The backbone of any sustainable supply chain relies on an effective and reliable transport network. Transport networks are clearly the lifeline that drives economic engines. Therefore it's pretty easy to deduce that the transportation sector needs to be a well oiled, highly efficient and highly productive system. An interesting research study that I recently read studied the importance and criticality of transport systems in supply chains and just as importantly how "sustainability" plays a vital role in planning and execution. From the study, the authors concluded.

"Without transportation, inputs to production processes do not arrive, nor can finished goods reach their destinations. In today's globalized economy, inputs to production processes may lie continents away from assembly points and consumption locations, further emphasizing the critical infrastructure of transportation in product supply chains."

"Indeed, companies are increasingly being held accountable not only for their own performance in terms of environmental accountability, but also for that of their suppliers, subcontractors, joint venture partners, distribution outlets and, ultimately, even for the disposal of their products. Consequently, poor environmental performance at any stage of the supply chain may damage the most important asset that a company has, which is its reputation.

In some cases, companies that don't measure their carbon emissions are finding themselves shut out of contract opportunities. Research has shown that trucking, rail, marine and air modes of transport all have their up and down sides and it's best to look at point to point options that will result in lower energy/fuel costs, use of modes that use cleaner fuels (LNG, ultra low sulphur diesel), and generate fewer greenhouse gas emissions (use of larger ships that employ more efficient equipment or operational practices). To that end, the transportation and logistics sector has been proactively looking at ways to improve efficiency, while simultaneously reducing environmental footprints associated with moving goods.

This chapter cannot get into the full range of transport avoidance, operational and technological changes that can be implemented to reduce the environmental footprint associated with moving goods however you can find these information in one of my books "Supply Chain management for Competitive Advantage". However, as an example, Freightliner Trucks addressed the issue of fuel savings by focusing on more efficient aerodynamics. The aerodynamic features to the company's Cascadia truck result in 7.8 percent to 22 percent less drag than other aerodynamic tractors, resulting in annual fuel savings of $900 to $2,750 per truck. Translate that into carbon emission reductions and the numbers would be enormous.

Meantime, there are also a number of tools that are available to assess Green House Gas (GHG) emissions and other environmental attributes associated with supply and transport, to allow you to accurately capture data and measure the true value of your supply chain and that is where some companies offer ways to measure and quantify what are called Scope 3 (indirect) carbon emissions. These companies can help to ship, track, measure carbon emissions.

The transportation sector makes great leaps in addressing its environmental footprint, so if you have not already started exploring your own environmental footprint, it's a great time to start leading the way or risk being a laggard.

Chapter 11: A Sustainable Transport Plan

The Sustainable Transport Plan 2006 -2016 sets out a 10-year programme of projects and actions that will help Aucklanders make safer and more sustainable travel choices. Most of the planning in transport goes into infrastructure (roads, railways and bus stations) and services (buses, trains and ferries). The third component of the transport system is people; specifically the transport choices of individuals, and of their schools, workplaces and neighbourhoods.

Understanding and influencing these choices is an essential component of Auckland's overall plans to achieve a world-class transport system. Sustainable Transport is defined in this plan as; working with people and their communities to improve travel opportunities and to encourage people to make fewer car journeys.

Preparing and developing the plan

The Sustainable Transport Plan has been prepared with input from the Regional Walking and Cycling Group, the Regional Stakeholder Group for School Travel Plans, the National Travel Behaviour Change Group, RoadSafe Auckland and many other groups, agencies and individuals. Detailed submissions were received from 47 organisations and individuals in the consultation phase, and these submissions resulted in significant changes to the plan.

This Sustainable Transport Plan sets out the actions needed to deliver the Sustainable Transport component of the Regional Land Transport Strategy, to be implemented over 10 years, to 2016.

Purpose of the plan

The plan aims to integrate sustainable transport activities with each other and with planned improvement to infrastructure and services. Walking, cycling, passenger transport and vehicle networks are all part of an overall transport system, and need to operate in an integrated way and to improve in response to local needs.

Using the plan

Getting this to happen, in a way that contributes to regional and national goals, requires working across multiple agencies and developing new ways of sharing costs, managing risk, and evaluating success. Auckland Transport (formerly Auckland Regional Transport Authority) will work in partnership with all transport agencies in Auckland to deliver the activities in this plan.

Reviewing the plan

The plan will be reviewed every three years to ensure it remains relevant and responsive to new ideas to improve sustainable transport for Auckland, and aligned to the Regional Land Transport Strategy.

Chapter 12: Green Supply Chain Brings Competitive Advantages

Well, can the economic tides be turning? In San Diego, they had a saying: "It takes a long time to carefully turn an aircraft carrier around".

Some key findings of note from a supply chain perspective:

1. Over 58 percent of the supply chain managers say their main business driver for 2015 is "Meeting (changing) customer requirements". (Well, I guess that is a no-brainer, as a successful business should be nimble and always responsive to customers' needs to succeed in the marketplace)

2. More than 50 percent of the participating companies indicate they will start up or continue with operational excellence / LEAN. Another obvious direction; reduces waste, optimize resources. This should translate into bigger profits and competitive position.

3. Sustainability is the second most important business driver for 2015 up 20 percent over this time last year however, the survey results suggest that this has not yet directly translated into a significant increase in supply chain sustainability projects. Well, remember that aircraft carrier quote that I just mentioned.

These findings really suggest that while the road to recovery is long, that much foundational work remains. But the trend from survival to revival is in play now.

Perhaps the biggest take-away from this report is the increasing emphasis of supply chain management in creating the proper ingredients of a successful business strategy and coincidentally, the concept of a Green Supply Chain is gaining interest among operations practitioners as a sustainable and profitable undertaking. A Green Supply Chain can be thought of as a supply chain that has integrated environmental thinking into core operations from material sourcing through product design, manufacturing, distribution, delivery, and end-of-life recycling.

The implementation of Green Supply Chain initiatives has evolved from strictly a compliance issue into a means of generating value. Traditionally, companies incorporating green projects have focused solely on cost avoidance by assuring compliance, minimizing risk, maintaining health, and protecting the environment. In the emerging value-creation model, implementing green initiatives along a company's supply chain can raise productivity, enhance customer and supplier relations, support innovation, and enable growth. The Green Supply Chain is no longer exclusively about green issues, but also about generating efficiencies and cost containment. As organizations restructure to reduce their company's environmental footprint, supply chains have increasingly become a key area of focus. Improvements in transportation efficiency, operations, raw material selection and packaging are all topping the list of "green" supply chain initiatives.

Green Supply Chains enable organizations to:
1. Specialize and concentrate manufacturing efforts in a way that manages environmental risks and costs of compliance with existing or new regulations.
2. Improve product, process, and supply quality and productivity.
3. Make innovative decisions that respond to "green economy" requirements.
4. Gain access to key markets through ISO 14001 registration or other certifications.
5. Improve or create brand differentiation and customer loyalty by offering unique capabilities to address environmental related requirements and expectations.
6. Reduce customer pressure and even gain preferred status.

The ISO 14001 Certification/Supply Chain Nexus

Over the past several years, studies have been performed worldwide comparing ISO 14001-2004 and its value in development of green supply chains.

One recent study found that more than 75% of manufacturing executives surveyed had ISO 14001 certification or were in process in order to enhance their competitive supply chain position.

Companies that are already ISO 14001 certified are 40% more likely to assess their suppliers' environmental performance and 50% more likely to require that their suppliers undertake specific environmental practices.

Preference in market share is often given to suppliers that have attained ISO 14001-certification.

Consumer preferences are increasingly important drivers for many companies to improve their supply chain environmental activities.

Procurement officers increasingly use ISO 14001 certification as a required vendor qualification.

Suppliers without an environmental management system will feel increasing pressure to modify their practices or risk losing customers, and will be subject to higher costs for licenses, inspections and insurance.

Questions and issues to consider when developing your Supply Chain/Value Network.

Will the service provider enhance the cause of sustainability both upstream (i.e., primary customer/end customer) and downstream (i.e., all tiers of supply base, including logistics service providers)?

Will some relationships drive significant redesign of the supply chain, including product innovations and modifications (e.g., collaborative development of decomposable packaging material?

Is your supply chain implementing progressive environmental management systems to manage their environmental footprint?

Establish a more cohesive collaborative model in transport, warehousing and distribution that will drive efficiencies up and incremental costs down, while reducing environmental impacts throughout the supply chain.

Environmentally responsible procurement, in alignment with your company's environmental sustainability values, is critical for organizations that desire to manage their environmental risk and maintain a competitive advantage.

Not only does this mean that businesses must choose their suppliers well, they also have to ensure that suppliers comply with the standards they claim to meet.

Chapter 13: Sustainable Supply Chains

While many companies targeted immediate supply chain value, some of the more well known leading organizations continuously assess the health of their supply base. It is this critical sustainable supply chain management activity that provides an opportunity to immediately determine risk and ultimately weather the changes in the market.

Using the impacts of the Japanese tsunami on business sustainability within the supply chain as an example, we could argues that traditional efficiency programs may not be sustainable in isolation. Instead, companies should selectively assess their lean improvement efforts and align them with a broader set of stakeholders.

Assess where you can and cannot strive for greater efficiency within your organization. When implementing a continuous improvement initiative, make sure you work with your product development team to discover non-value-added activities in your current workflow.

Our experience reveal that over the past few years business sustainability action has resulted in strategic sourcing and procurement guidelines seeking to align suppliers with defined business sustainability strategies. These efforts have transformed traditional supply chain efficiency programs to include a broader set of factors, which affect the long-term stability of the company.

Moving forward, we subscribes to the idea that the next generation of sustainable supply chain management will continue to emerge. Focused on a new level of responsibility across all supply chain activities, efforts in 2014 will drive deeper into category value by addressing more pointed supply questions. As such, we explore emerging trends in the sustainable supply chain through the real examples of this progress already taking form.

Five Lessons from Wal-Mart on Making Supplier Scorecards Work for You. We examine how Wal-Mart manages an extremely wide range of suppliers; from the extremely small local supplier to the technically advanced mega corporation. Some of the key takeaways from their supplier sustainability program include.

1. Management buy-in at key levels: Buy-in is obviously essential at the CEO and senior management level, but equally critical is support at the management level one or two levels above buyers, such as senior and division merchandising managers.

2. Incentives: Financial rewards and recognition send clear signals to the organization about the importance of sustainability.

3. Right training, tools, and support: Awareness and motivation are only part of the mix needed for success. The correct type and level of training, tools, and support are vital.

4. Make the scorecard program a part of existing business processes: Supplier sustainability programs need to be integrated with existing business processes, such as the joint business planning and annual supplier reviews, and core merchandising processes, such as category strategy development and line reviews.

5. Celebrate, celebrate, celebrate: Informally and repeatedly celebrating success is critical throughout the organization. Managers and buyers share short success stories in meetings, during internal town hall gatherings, and on company blogs.

Sustainability is a collaborative process. There is only so much that any one business can do about acting responsibly on behalf of the environment or society without involving its supply chain and holding it accountable.

Chapter 14: Motivating Suppliers to Meet Sustainable Sourcing Requirements

Are you old enough to remember the opening lines of the Buffalo Springfield song For What it's Worth? "There is something happening here/What it is ain't exactly clear. There is a man with a gun over there. Telling me I got to beware". I am thinking there is a green supply chain revolution in play, just as there was political unrest and turbulence of the mid to late 1960's from which this song originated. I thinks Wal-Mart may be "the Man", but are they really holding a gun to suppliers? I'm not so sure.

Wal-Mart's efforts internally to establish its sustainability index continue to slowly progress along (I still predict a 2-3 year process before anything tangible emerges). But, the company is as I predicted, changing the rules in how sustainability is felt up and down the supply chain; mostly for good. Many companies in the retail and electronics sectors, such as Proctor and Gamble and IBM have most notably stepped up to the plate, but many others are learning from Wal-Mart's green supply as well. So how is this "cat herding" happening at such a rapid pace and what are the key issues being driven through the 'value chain'. Is this just a matter of keeping up with the next guy?

First - the drivers. There are a number of factors and issues, both internal and external that can be attributed to this hot phenomenon in the supply chain pace. A recent research indicates that, sustainability was clearly a driving topic in supply chain management, ranking behind only three factors; improving customer service, reducing supply chain risk and managing and optimizing an extended supply chain network. The same study found that several factors were driving the greening of the supply chain across a number of vertical markets, notably.

1. Lost sales (projected to be in the billions of dollars) because products in the supply chain were not "green" enough.
2. Increased energy and transportation costs (accounting to over 50% of the cost increases).
3. Damage to reputation.
4. Supply disruptions.

In response, Wal-Mart and other major retail and industry giants are driving upstream and downstream performance based changes, designed to reduce suppliers environmental footprints and focused on several key areas; such as energy management, fuel cost containment, carbon emissions, water use and waste generation. New issues also factoring into the mix include green chemistry and management of restricted materials, depending on the geographic reach of global markets served.

To that end suppliers, from Tier One on down through the chain are responding to varying degrees and the early results appear favourable. Companies like Wal-Mart, P&G and Johnson and Johnson are showing marked reductions in most of the key metrics that they have been focused on, with much of the credit due to those suppliers who have found business sense in sustainability.

Now to that you may say that suppliers are goaded, cajoled, forced, strongly encouraged, or perhaps threatened to comply, or else risk losing millions in contracts. Actually, what I am seeing with the likes of Miller, IBM, Hewlett Packard and others continues to be more of the carrot and less of the stick; more collaboration and performance based incentives coupled with onsite verification that is all good because it encourages accountability.

Chapter 15: Sustainable Procurement

To paraphrase a timeless Bob Dylan song, "The Times They Are A' Changing'" is no understatement. You can read the details from across the globe in the news every day and are rapidly happening simultaneously on political, economic and social levels and businesses are also making radical changes in the sustainability and corporate social responsibility (CSR) frontier.

"Then you better start swimming' Or you'll sink like a stone. For the times they are a-changing'."- Dylan

One area that appears to be in movement is Procurement. You know, those folks on the third floor in the back that order stuff? Well, wrong! I have maintained that the heart of a sustainable supply chain runs through its procurement function. That is because every product; every single purchase has a hidden human health, environmental and social impact along the entire supply chain. One of my books (Lean Procurement and Supply Chain management - Key to Reducing Costs and Improving Profitability) discussed how the procurement function is a vital cog in product value chain. Purchasing staff are the "gatekeepers" that can access powerful tools and serve as a bridge between supplier and customer to assure that sustainability and CSR issues are taken into account during purchasing decisions. 2014 was a watershed year for sustainability initiatives and supply chain management and I predicted that 2015 would see greater progress.

So I was incredibly excited when I recently got my hands on a relatively new white paper from Ariba, entitled "VISION 2020 - Ideas for Procurement in 2020 by Industry-Leading Procurement Executives". According to the conveners of the document, the "objective of the effort initiated in 2010 is to initiate a dialogue on the future of procurement and to create a roadmap for how to get there." For that, they connected with leading practitioners and executives from around the world and across a variety of sectors to share their ideas, best practices and to read the tea leaves as to where procurement might be in 10 years.

While the initial report laid out some pretty intriguing and widely varying trends and predictions about the state of procurement in the corporate function, I was unfulfilled. I was all ready to read about how the emergence of sustainability in the marketplace was going to drive procurement decisions. I expected to hear how top flight companies around the world were collaborating with their supply chain, implementing staff training on 'green purchasing' practices, and implementing sustainability driven supplier audits and ratings scorecards.

Boy, was I wrong! Only ONE mention of the word "sustainability" (thank you Dr. Heinz Schaeffer, Chief Procurement Officer, Northern and Central Eastern Europe for AXA), and no mentions of "responsible sourcing", "green supply chain" or "sustainable sourcing". I would have expected more from chief procurement representatives from the likes of KeyBank, Maersk, Sodexho, and former execs from Hewlett Packard, General Motors, and DuPont. Most of these companies are generally considered leaders in the sustainability space. So why would there be a disconnect between what companies are doing in design, manufacturing and product life cycle management and the procurement function?

Before we go too far, it's helpful to define what "sustainable procurement" is. While there is no singular definition for it, I like the definition offered up by the UK-based Chartered Institute of Procurement and Supply (CIPS). CIPS definition is "a process whereby organisations meet their needs for goods, services, works and utilities in a way that achieves value for money on a whole life basis in terms of generating benefits not only to the organisation, but also to society and the economy, whilst minimising damage to the environment.". What CIPS defines as 'whole life basis' is that "sustainable procurement should consider the environmental, social and economic consequences of design; non-renewable material use; manufacture and production methods; logistics; service delivery; use; operation; maintenance; reuse; recycling options; disposal; and suppliers capabilities to address these consequences throughout the supply chain".

It's a good thing that the authors from Ariba stated that "The 2020 Vision report is intended not as an end, but rather as a point of departure for much discussion and debate around where procurement can and should be setting its sights for the year 2020 and beyond. In fact, Ariba invites readers to "join the debate and to extend the discussion with new ideas by joining the conversation. I have and I hope you will too.

Key Findings of Interest

Ariba report identified six key trending areas and takeaways among the participants who have weighed in so far, namely.

1. Procurement devolves: With spend management requirements shrinking, companies are being forced to optimize what resources they have and make better informed decisions. More work at the business line level will occur, possible eliminating the central procurement function entirely. Money and metrics will drive most decisions as companies face leaner profit margins. There will be a need to engage end customers more and more and leverage relationships.

2. The new supply management emerges: Some traditional sourcing functions may become outsourced. Strategy "will tie directly to an enterprise's end customers and it will be more cognizant of the diversity of desires and requirements within the customer base".

3. Skill sets change: The Chief Procurement Officer and staff must have broader skills that allow them to not only create opportunities for revenue enhancement internally and optimized "spend", but also be more in touch with end customer values-driven needs. Procurement staff need to be tuned into multiple tiers of the supply chain, dive deep "inside the supply chain and bring issues forward to the designers within individual companies".

4. Instantaneous intelligence arrives: Market pricing will become more transparent; the Cloud forces transparency to some degree. Companies will have to rapidly extract innovation and other value from supplier bases, and build exclusive commercial relationships with leading suppliers that share both risks and rewards.

5. Collaboration reigns: There will be as the report notes a "big emphasis on driving and taking innovation from the supply base... the supply role will be less 'person who brings innovation in' and more 'person who assembles innovation communities and gets out of the way'. Suppliers are being asked more often to participate in early design and product development as a way to leverage risk and control overall product life cycle management risks.

6. Risk management capacity and demands soar: As companies are already realizing, effective procurement relies on response to risk management variables (financial, ethical, and operational performance). Companies must create 360-degree performance ratings and provide greater transparency into market dynamics, potential supply disruptions, and supplier capabilities. There will be a big expansion in the kinds of risks companies address in their supply chains, considering, for example, such things as suppliers' sustainability, social responsibility.

Now if I read in between the lines, I can easily pluck out a number of key procurement trends from the 2020 report that transfer well to sustainability and responsible sourcing. Risk Management. Collaboration. Design phase (life cycle) engagement of multi-tiered suppliers. Key performance metrics. Responding to consumer demands. Supplier performance ratings.

One takeaway for me appears that there may be a disconnect still between the procurement function and other functions within organizations. So is the procurement function still operating in obscurity in most organizations? It all depends who you talk to but also on your skill at reading the tea leaves.

Rest assured that compared to only a few years ago, more companies that are seeking to manage the life cycle environmental impact of their products from design and acquisition of materials through the entire production, distribution and end of life management. They are finding sustainable procurement to be a valuable tool to quantify and compare a product or component's lifetime environmental and social impact early on in a products value chain while positioning the company for smart growth in a rebounding economy. We may be at a sustainable procurement tipping point.

Chapter 16: Game On to Drive Sustainable Procurement

On the heels of the Ariba effort comes a promising benchmark report released by HEC-Paris and Ecovadis. Entitled "Sustainable Procurement: back to management!" has risen to rescue and tempered my fears of devolving sustainable procurement. In fact, the report may suggest a positive "tipping point" in favour of sustainable procurement. The efforts behind the 2011 edition of the HEC/EcoVadis Sustainable Procurement Benchmark were carried out between the fall of 2010 and early 2011. This benchmarking process started in 2003.

The objective of the benchmark is to provide a snapshot on what is trending in the area of Sustainable Procurement practices. According to the authors, the following overarching questions were explored:
1. How has the vision of the Chief Procurement Officers (CPOs) evolved?
2. What tools and initiatives seem to be the most effective over time to drive changes?
3. How is Sustainable Procurement progress measured?
4. What are the remaining challenges faced by most Procurement organizations?

The study identified three main drivers behind Sustainable Procurement initiatives; Risk Management, Value Creation, and Cost Reduction.

These findings mirror some of the trending areas and critical issues identified in the Ariba report. HEC and Ecovadis suggested that these three drivers' shows that many organizations are now facing new expectations in terms of Corporate Social Responsibility and Sustainability from the Procurement Departments of their clients and, suggest that having a sustainable procurement program in place can become a competitive advantage.

Sustainable Procurement Remains High on Executives Agenda

92% of the surveyed Companies consider Sustainable Procurement a "critical" or "important" initiative, even though for the 1st time this year, "Risk Management" took over as a priority initiative.

The major progress made in 2014 is on the support from the Top Management (+24%) thus demonstrating that Sustainable Procurement is attracting more and more interest from Executive Committees, and significant progress was made in implementation of tools and organizational changes.

Significant organizational changes have been implemented; 45% of companies already have "dedicated teams" and 57% report having trained a majority of procurement staff on Sustainability.

Whereas in 2007 only a third of companies were using formalized methodologies for assessing their suppliers' sustainability performance, in 2014 two-thirds of them are now implementing dedicated tools (either internal or leveraging 3rd parties).

Finally 92% of companies have increased (56%) or maintained (36%) their budgets related to Sustainable Procurement, which should yield more changes in the future years.

Tools for Sustainable Procurement on the Rise

The HEC/Ecovadis study found that basic tools such as "Suppliers Code of Conduct", "CSR contract clauses" and "Suppliers self-assessment" were now the rule rather than the exception among companies surveyed by a ratio of 2 to 1, but interestingly were still found to limited value in terms of risk management. What I found encouraging was that the study found maturation in the types of tools used, including "Supplier Audits" and "Supplier CSR information databases". This type of work has clearly been evident in what we have reported in the past, especially among multi-national companies with contract manufacturing operations in developing economies (like China, India and Brazil). These advanced tools offered more opportunities for suppliers to engage directly with buyers, allow for data verification, and offer direct recommendations for supplier CSR and sustainability improvement. Over half of the companies surveyed had advanced to this next level. When asked what the most effective uses of resources were in developing a Sustainable Procurement Program, respondents mentioned.
1. Top level support.
2. Creation of cross functional teams.

3. Training, as key success ingredients.

All three of these success factors had shown substantial improvement over the past several benchmark cycles, according to the study.

Sustainable Procurement Creates Value

Payback from most green procurement activities is huge. Companies surveyed were able to benefit quickly from risk management reduction and potential revenue growth opportunities, due in part to sustainable procurement. The study also found that there were additional 'value creation' opportunities that could be realized if procurement departments collaborated more closely with the marketing and R&D departments upstream on the projects.

Also, a study in 2009 by a company named BrainNet (Green and Sustainable Procurement: Drivers and Approaches") looked at sustainable procurement and value creation and found that "procurement with an ecological and social conscience is not a cost factor, but a value factor. Companies that pursue a consistent approach to green and sustainable procurement receive an above-average return on capital deployed." The study produced what they describe as an "evolution curve for sustainable procurement" that describes the maturity of various approaches of sustainable procurement. This curve compares well with the most recent EcoVadis/HEC findings and suggests that there may be a widening gap between leaders and laggards.

Sustainable 'green' procurement embraces a holistic approach, one that encompasses organization, people, process, and technology to create greater product value along the entire supply chain. This type of value creation can managed by establishing firm triple bottom line based metrics from upstream suppliers to downstream users and using the procurement function to support product and process innovation and accounting for total cost of ownership (TCO).

What's Next?

According to the a recent HEC/EcoVadis benchmark report, it is clear that new green and social business models depend upon innovation, and a gap still among many organizations to implement a truly Sustainable Procurement vision.

The HEC/Ecovadis report suggests that when implementing Sustainable Procurement practices, a three phase process can get the ball rolling, starting first by orienting and energizing the procurement function through.

1. Communication activities: Building awareness among employees regarding the approaching change, the benefits and the steps to be implemented.

2. Training and Performance support: Ensuring that the initiative is being understood among those who are to execute the change or be part of it, and leading to buy-in of the key stakeholders.

3. Rewards and recognition: Ensuring that employees and suppliers who embrace change are properly recognized and rewarded. This final step is when implementation is not only measured, but also celebrated.

I am going to say it again and again. All sustainable business roads lead through the procurement function. The procurement function is the perfect nexus and a critical organizational player that touches product designers, engineers, multiple tiers of suppliers and subcontractors, manufacturing operations, logistical warehousing and distribution and the end users. Yes indeed, things are looking up for sustainable procurement...it's 'game on'.

Chapter 17: Future-Proofing Supply Chains With Sustainable Procurement

Sustainable sourcing is turning out as the most important function of supply chain management and the reason why sustainable sourcing may very well become the single most important aspect of supply chain management, is because of how it grounds itself in future-proofing supply chains.

What it means to future-proof supply chains.

Future-proofing supply chains means protecting your organizations performance in the future against a multitude of potential problems. Whether the future involves loss of access to key production ingredients or logistics providers, future-proofing attempts to mitigate potential dangers.

Trending in the environment towards future-proofing supply chains is sustainable procurement. Sustainable procurement has proven that it improves efficiency, effectiveness, and transparency for organizations. Furthermore, sustainable sourcing not only improves employee and supplier relationships, it also improves an organizations reputation amongst their community and their customers. Lastly sustainable sourcing can dismantle poor waste disposal practices and reduce energy waste, improving company spend management.

Beyond the business case towards sustainable sourcing, ecologically friendly procurement practices are necessary for maintaining our environment.

Benefits of sustainable sourcing.

Amongst the numerous benefits sustainable supply chain management (SSCM) can provide; the more important benefits includes competing with a green strategy and improving market competitiveness.

Green strategy involves reducing waste management costs, trading ethically to attract ethically-conscious customers, and reducing energy resource waste.

Improving market competitiveness involves reducing procurement delays by adequately projecting future sourcing activities against potential environmental hazards, government regulations, shift in customer attitudes towards environmental concerns, and changes in supplier relationships.

In fact, sustainable sourcing practices can enhance buyer-supplier relationships through increased transparency.

Why E-procurement is often considered the best step towards SSCM.

Before we get into why most organizations are choosing to optimize their supply chain consider this single piece of info. "One manual purchase order can cost a company as much as $100 to process, even if the purchase order was issued to buy a $25 part.

The cost of issuing manual purchase orders add up when considering the time spent sending documents back and forth between approvers, buyers, and receivers. E-procurement systems have become significantly more popular as a means of removing the tedious travel time of documents.

There are three significant ways that e-procurement improves the purchasing process.

1. Goodbye paperwork. E-procurement systems eliminate paperwork by generating electronic purchase orders which can be immediately routed to the appropriate person. Resulting in less paper cost from purchasing paper and disposing of it, as well as less time spent between key personal.

2. Commitment to vendors. E-procurement systems can be used as a means of compelling end users to only purchase from company approved vendors. This commitment to approved vendors is one way of mitigating the possibility of corporate fraud within the organization. An example of such would include a fake vendor where funds can be funnelled into.

3. Spend visibility. Although spend visibility may sound like a generic buzzword, spend visibility is an integral function of e-procurement systems. Spend visibility allows managers full control and their company and department spending.

Chapter 18: Sustainable Packaging

The pea pod is possibly the greatest sustainable packaging design nature can provide. It packs a lot in a small space, efficiently uses the minimum amount of resources and best of all its compostable; well sort of unless I eat it!

Like the simple pea pod, few sustainability attributes in a supply chain come together across the value chain than packaging. Packaging and repackaging is ubiquitous along every step of the chain, from product design, prototyping, procurement production, distribution, consumer end use and post consumer end-of-life management. The more parts that are in use in making of a product, and steps along the way to deliver the parts, the greater the packaging (and hence environmental footprint) involved along that chain and for every packaged part that comes from someplace else to make a product, a similar carbon, energy and resource use can be measured.

That is why sustainable practices in packaging are so important in driving supply chain efficiency and why innovation in the 'green' packaging sector has been "red hot" in the past several years. A study by Accenture found that retailers can realize a 3 percent to 5 percent supply chain cost savings via green packaging initiatives. So if you extrapolate that type of savings out across multiple tiers of supply chain activity, where packaging is the common denominator, the efficiencies and savings can rack up quickly.

A new report from a research organization finds that because of a variety of drivers such as carbon emissions, extended producer responsibility and waste reduction targets plus advanced packaging technologies, the sustainable and green packaging market is expected to reach $127 billion in 2015. The report shows varying degrees of growth from developed to developing nations; however what is striking is that the growth trend is weathering the slumping global economy and higher production costs.

Sustainable packaging solutions deliver around two colours; black (deliver reduced costs) and green (reduce environmental impacts). Sustainable packaging relies on best engineering, energy management, materials science and life cycle thinking to minimize the environmental impact of a product through its lifecycle. Given the past decade or so of science and engineering work around sustainable packaging, there are some discovered and tested attributes, such as.
1. Reducing packaging and maximizing the use of renewable or reusable materials.
2. Using lighter weight, less toxic or other materials which reduce negative end-of-life impacts.
3. Demonstrating compliance with regulations regarding hazardous chemicals and packaging and waste legislation (such as the European

Directive 94/62/EC on Packaging and Packaging Waste).
4. Optimizing material usage including product-to-package ratios.
5. Using materials which are from certified, responsibly managed forests.
6. Meeting criteria for performance and cost (e.g., minimize product damage during transit).
7. Reducing the flow of solid waste to landfill.
8. Reducing the costs associated with packaging (i.e., logistics, storage, disposal, etc.)
9. Reducing CO2 emissions through reduced shipping loads

Best in Class Examples

I have seen companies stress the importance of the 6 R's of sustainable packaging (refill, reduce, recycle, repurpose, renew, reuse; Wal-Mart's 7 R's of Sustainable Packaging (Remove Packaging, Reduce Packaging, Reuse Packaging, Renew(able), Recycle(able), Revenue (economic benefits), and Read (education); and even the 10 R's eco-strategy (Replenish, Reduce, Re-explore, Replace, Reconsider, Review, Recall, Redeem, Register and Reinforce).

Associations are stepping up to the plate as well as manufacturers in a variety of consumer product markets. The Grocery Manufacturers Association (GMA) announced the results of survey a research that indicated elimination of more than 1.5 billion pounds (800 million pounds of plastic and more than 500 million pounds of paper) since 2005, and another

2.5 billion pounds are expected to be avoided by 2020. Over 180 packaging initiatives were identified and evaluated. The GMA estimated that the reduction would be equal to a 19 percent reduction of reporting companies total average U.S. packaging weight.

In the fast moving consumer goods category Coca Cola's packaging efficiency efforts just in 2009 avoided the use of approximately 85,000 metric tons of primary packaging, resulting in an estimated cost savings of more than $100 million. The company rolled out of short-height bottle closures, reducing material use, implemented traditional packaging material light weighting; and used more recycled materials in packaging production. At the end consumer point, the company has also supported the direct recovery of 36% of the bottles and cans placed into the market by the Coca-Cola system and continues to work with distributors on increasing recovery efforts.

In the electronics space, Dell Computer committed in 2008 to reduce cost by $8 million and quantity by 20 million pounds of packaging by 2015 centred around three themes (cube, content, curb).
1. Shrinking packaging volume by 10 percent (cube).
2. Increasing to 40 percent, the amount of recycled content in packaging (content).
3. Increasing to 75 percent, the amount of material in packaging to be curbside recyclable (curb).

As an example, Dell wanted to find a greener, more cost efficient way to package its computers by eliminating foams, corrugated and moulded paper pulp. The solution was sustainably sourced bamboo packaging certified by the Forest Stewardship Council. So far, Dells efforts have resulted in eliminating over 8.7 million pounds of packaging, and they have nearly met their recycled content goal.

Perhaps most significantly, Wal-Mart took a huge step in 2007 to seek supplier conformance around packaging. Since then, despite the initial uproar, there has been an uptick in design and innovative product activity by thousands of key suppliers in response to the mega-retailers challenge. By reducing packaging in the Wal-Mart supply chain by just five percent by 2015, that would prevent 660,000 tons of carbon dioxide from entering the atmosphere, keeping 200,000 trucks off the road every year (that is a green attribute) and save the company more than $3.4 billion (a black attribute). Wal-Mart's bottom line was to put more products on its shelves in the same space, and also recognized the sustainability attributes that change would make. They also knew that most consumers (me included) just despise excess packaging. Here are two examples of Wal-Mart supplier efforts from a small and large supplier:

Alpha Packaging: the company has a new bottle design for Gumout Fuel Injection Cleaner. The company concentrated the product and switched from PVC bottles (which are not recyclable) to much smaller bottles made from PET (which is recyclable

and has 30% post-consumer recycled content). This led to.
1. Reduced product weight by up to 51%.
2. Capability to transport a truck filled with new 6 oz products (formerly 12 oz) equating to 153,600 bottles as opposed to 61,000 originally.

General Mills: the company took a novel approach and they looked at the product first. They straightened its Hamburger Helper noodles, meaning the product could lie flatter in the box. This, in turn, allowed General Mills to reduce the size of those boxes. According to the company, that effort saved nearly 900,000 pounds of paper fiber annually. The company effort also managed to reduce greenhouse gas emissions by 11 percent, took 500 trucks off the road and increased the amount of product Wal-Mart shelves by 20 percent. Win-Win-Win. For the environment, for manufacturers and suppliers, and for consumers.

Full Circle Collaboration is Vital to Drive Sustainable Packaging

What makes sustainable packaging compelling is that it's one of the key elements of a product that consumers can see, touch and feel. Over packaging or improper packaging can produce high reaction levels, right? (remember noisy Sun Chips compostable bag dust up?) What was a scary statistic is that brand owners and retailers may have direct control over as little as 5 percent of the environmental impacts of packaging and only indirect control over the other 95 percent. On the other hand another study conducted by a market research firm showed U.S. consumers surveyed; 49% felt that packaging design has a medium or high level of influence over their choice of food and drink products. Just as there are challenges to drive consumer acceptance of more sustainable types of package designs (especially aesthetics), there are equally challenging design factors (such as package strength, permeability, and other physical factors that may compromise product integrity during shipment.

Opportunities to Leverage the Supply Chain from Design to Post Consumer Package management.

High performing manufacturing companies are clearly using sustainable packaging design and manufacturing as a way to lever efficiencies through the product value chain. Companies are finding that using less complex packaging helps cut sourcing, energy production and distribution and fuel costs across the supply chain. The glory days of corrugated packaging as the one stop solution are being replaced with reusable packaging options. Also, reducing the consumption of raw materials, carbon emissions and waste generation reduces manufacturing costs. Since disposal by consumers is one of the largest waste streams in the supply chain, using less packaging of direct-to-consumer shipments also offers great opportunities for supply chain optimization. The previously mentioned research report recommends that through route planning and sourcing software, "collaboration across the companies in the supply chain is necessary to maximize freight utilization. In particular, retailers need to proactively encourage vendors to provide pallet or "trailer feet" specifications for collecting shipments; retailer's planners can determine the optimum transportation mode and look for multi-stop opportunities."

Optimized Supply Chain

There are suggestions that opportunities to reduce the packaging/un-packaging cycle by addressing the product life-cycle and optimized material use. Through ongoing recycling and the use of alternative materials throughout the product value chain, opportunities are created to reduce the volume of packaging waste. Also, take back programs create a two-way transportation flow, with reusable packaging materials being sent back up the supply chain rather than to a landfill.

Some final pointers to consider when designing packaging and using the supply chain to drive sustainability.
1. Source alternative sustainable packaging materials; the innovative options are ample.
2. Evaluate product life-cycle impacts as a way to discover design options that could lead to less packaging.
3. Anticipate the total energy and resource use over an entire products package life.
4. Evaluate materials disposal and post consumer end-of-product life opportunities.
5. Design products for efficient transport.
6. Schedule and optimize transportation networks.
7. Collaborate, Collaborate, Collaborate!

Chapter 19: Designing Sustainable Products and Services

At this stage executives start waking up to the fact that a sizable number of consumers prefer eco-friendly offerings, and that their businesses can score over rivals by being the first to redesign existing products or develop new ones. In order to identify product innovation priorities, enterprises have to use competencies and tools they acquired at earlier stages of their evolution.

Companies are often startled to discover which products are unfriendly to the environment. When Procter & Gamble, for example, conducted life-cycle assessments to calculate the amount of energy needed to use its products, it found that detergents can make U.S. households energy guzzlers. They spend 3% of their annual electricity budgets to heat water for washing clothes. If they switched to cold-water washing, P&G reckoned, they would consume 80 billion fewer kilowatt-hours of electricity and emit 34 million fewer tons of carbon dioxide. That is why the company made the development of cold-water detergents a priority. In 2005 P&G launched Tide Coldwater in the United States and Ariel Cool Clean in Europe. The trend has caught on more in Europe than in the United States. By 2008, 21% of British households were washing in cold water, up from 2% in 2002; in Holland the number shot up from 5% to 52% of households. During the recent recession P&G

has continued to promote cold-water products, emphasizing their lower energy costs and compact packaging. If cold-water washing catches on worldwide, P&G will be able to cash in on the trend.

Companies are often startled to discover which products are unfriendly to the environment.

Likewise, we were surprised to learn that household cleaning products are the second biggest environmental concern after automobiles in the United States. Our market research also showed that 15% of consumers treat health and sustainability as major criteria when making purchase decisions, and 25% to 35% take environmental benefits into consideration.

In 2008 rumour has it that Clorox became the first mainstream consumer products company to launch a line of non-synthetic cleaning products. It spent three years and more than $20 million to develop the Green Works line, delaying the launch twice to ensure that all five original products performed as well as or better than conventional options in blind tests.

Clorox had to tackle several marketing issues before launching Green Works. It decided to charge a 15% to 25% premium over conventional cleaners to reflect the higher costs of raw materials. Green Works products are still cheaper than competing products, which carry a 25% to 50% mark-up over synthetic ones. After much discussion, the marketing team chose to put the Clorox logo on the Green Works line to signal that it performs as well as conventional

Clorox products. The company persuaded the Sierra Club; a leading environmental group in the United States to endorse Green Works. Although it sparked controversy among activists, this partnership strengthened Clorox's credentials, and in 2008 the company paid nearly $500,000 to the Sierra Club as its share of revenues from the line. Finally, Clorox struck special arrangements with retail chains such as Wal-Mart and Safeway to ensure that consumers could easily find Green Works products on shelves.

By the end of 2008 Green Works had grown the U.S. natural cleaners market by 100%, and Clorox enjoyed a 40% share of the $200 million market. Green Works sales weakened in the fourth quarter of 2008 because of the recession, but they rebounded in the first quarter of 2009. The tailwind has encouraged Clorox to launch more sustainable products: In January 2009 it introduced biodegradable cleaning wipes, and the following June it introduced non-synthetic detergents, where it will run into rival P&G.

To design sustainable products, companies have to understand consumer concerns and carefully examine product life cycles. They must learn to combine marketing skills with their expertise in scaling up raw-materials supplies and distribution. As they move into markets that lie beyond their traditional expertise, they have to team up with nongovernmental organizations. Smart companies like P&G and Clorox, which have continued to invest in eco-friendly products despite the recession, look beyond the public-relations benefits to hone competencies that will enable them to dominate markets tomorrow.

Chapter 20: Developing New Business Models

Most executives assume that creating a sustainable business model entails simply rethinking the customer value proposition and figuring out how to deliver a new one; however, successful models include novel ways of capturing revenues and delivering services in tandem with other companies. In 2008 FedEx came up with a novel business model by integrating the Kinko's chain of print shops that it had acquired in 2004 with its document-delivery business. Instead of shipping copies of a document from, say, Seattle to New York, FedEx now asks customers if they would like to electronically transfer the master copy to one of its offices in New York. It prints and binds the document at an outlet there and can deliver copies anywhere in the city the next morning. The customer gets more time to prepare the material, gains access to better-quality printing, and can choose from a wide range of document formats that Fed-Ex provides. The document travels most of the way electronically and only the last few miles in a truck. FedEx's costs shrink and its services become extremely eco-friendly.

Some companies have developed new models just by asking at different times what their business should be. That is what Waste Management, the $14 billion market leader in garbage disposal, did. Two years ago it estimated that some $9 billion worth of reusable materials might be found in the waste it carried to landfills each year. At about the same time, its customers, too, began to realize that they were throwing away money. Waste Management set up a unit, Green Squad, to generate value from waste. For instance, Green Squad has partnered with Sony in the United States to collect electronic waste that used to end up in landfills. Instead of being just a waste-trucking company, Waste Management is showing customers both how to recover value from waste and how to reduce waste.

New technologies provide start-ups with the ability to challenge conventional wisdom. Calera, a California start-up, has developed technology to extract carbon dioxide from industrial emissions and bubble it through seawater to manufacture cement. The process mimics that used by coral, which builds shells and reefs from the calcium and magnesium in seawater. If successful, Calera's technology will solve two problems. Removing emissions from power plants and other polluting enterprises, and minimizing emissions during cement production. The company's first cement plant is located in the Monterey Bay area, near the Moss Landing power plant, which emits 3.5 million tons of carbon dioxide annually. The key question is whether Calera's cement will be strong enough when produced in large quantities to rival conventional Portland cement. The company is toying with a radical business model. It will give away cement to customers while charging polluters a fee for removing their emissions. Calera's future is hard to predict, but its technology may well open an established industry and create a cleaner world.

Developing a new business model requires exploring alternatives to current ways of doing business as well as understanding how companies can meet customers' needs differently. Executives must learn to question existing models and to act entrepreneurially to develop new delivery mechanisms. As companies become more adept at this, the experience will lead them to the final stage of sustainable innovation, where the impact of a new product or process extends beyond a single market.

Chapter 21: Sustainable Supply Chain Alignment

In today's economy, there is increasing emphasis on supply chain management as being vital to a long-term successful business strategy. In the emerging "Green Economy", sustainable sourcing, environmentally responsible logistics, and 'Green' Supply Chain Management (GSCM) is gaining traction as an essential business management tool and a key leverage point in driving consumer product sustainability. GSCM integrates an environmental lens into core operational management; from material sourcing through product design, manufacturing, distribution, delivery, and end-of-life management.

Traditionally, organizations that implement sustainability-focused initiatives have focused solely on cost avoidance by assuring compliance, minimizing risk, maintaining health, and protecting the environment within their own "four walls". Studies have proven time and again that implementing sustainable sourcing initiatives along a company's entire supply chain, both upstream and downstream of the "four walls", can.
1. Raise productivity and contain costs.
2. Reduce product environmental footprint.
3. Enhance customer and supplier relations.
4. Support innovation.
5. Leverage brand trust.
6. Enable prosperous growth.

The biggest barrier to the success of a company's sustainable supply chain practices is a lack of leadership support.

About 30 percent of operations executives surveyed said their company has a documented supply chain sustainability strategy, but only 17 percent of managers and below agreed. As a result, mid-level management is not able to take the steps needed to drive meaningful change in the supply chain.

While disconnects about sustainable supply chain strategy may occur between the C-suite and mid-level management, 76 percent of operations professionals said their companies' focus on creating a more sustainable supply chain will increase over the next three years. Already, 43 percent of operations professionals attributed cost reduction to supply chain sustainability initiatives, while 35 percent reported improvements in their company's environmental impact. A quarter of all respondents reported improved customer satisfaction as a result of programs tied to improving supply chain sustainability.

The major barrier to supply chain sustainability cited in the survey was that leadership does not supply the mandate, incentives, and resources to turn sustainability strategies into action. Additional barriers reported by supply chain professionals included inadequate sustainability education and training, significant confusion about the scope and company goals on supply chain sustainability, and the perception that the impact on shareholder value for such practices is difficult to measure.

More than a third of professionals (38 percent) said that barriers to success included the ability to measure and monitor targets and goals. Another 40 percent believe employee performance measurement and incentives are not aligned to supply chain sustainability results.

Chapter 22: Watch Your Step

This week has been all about "R-I-S-K". Risk that my three flights to Nigeria will be on time. Risk that my luggage will accompany me. Risk that I will meet my taxi driver. Risk that he will be a safe driver, negotiating darkness and harrowing roads full of heavy trucks travelling between Abuja and Maiduguri. Risk that my digestive system can handle all the amazing foods I will sample while in Nigeria. Risk that my talk on integrated sustainability management systems will go off without a hitch.

Risk (noun): A situation involving exposure to danger.

Risk (verb): to expose to danger or loss.

Risk. We all live with risk and all are in position to control and influence its outcome. This week's conference was devoted to exploring risk in the workplace and its related effects on worker safety, health and environmental impact. Nigeria is the perfect place to explore this issue, because of all of the social, political, economic and workplace/environmental challenges that this special country has endured over generations. Throughout the 3 days conference I have become painfully aware of the risks that exist amid the beauty of the Nigeria Capital City of Abuja and Maiduguri.

This great place of beauty has seen terrorist attack from terrorist group Boko Haram. This is historic ground where people took incredible risks to protect what they believed in, and suffered enormous costs and joyous victories. I won't use this space to opine on that matter just to say that issues run deep and wounds take generations to heal. But all citizens of the Nigeria are trying their very best to level the playing field. But all along the way, all the players in this real life drama have had to manage risk.

Snakes!!

To illustrate how risk is all around us in the workplace and at home, they brought out the snakes...yes, snakes. Not the safe variety...I mean the pythons and puff adders. Through a safety company, the idea of "Snakes for Safety" was presented to a fascinated, but somewhat skittish audience of 400. The analogy is that puff adders are like accidents waiting to happen; they hide, camouflaged in the bush and only strike when you are right on top of them. By then the damage has been done, injury's result (and in the case of the puff adder, you have seven minutes to call a loved one and say goodbye!). Cobras on the other hand represent a hazard that is harmless when small, but if left unchecked, the hazards can grow to an unmanageable point when great harm can occur. Snakes. Risk. Managing the basics of health, safety and the environment (HSE) in developing economies like Nigeria is foremost in businesses minds and correctly so.

"There are risks and costs to every program of action. But they are far less than the risk and costs of comfortable inaction"- John F Kennedy

Continuous risk management process helps organizations understand, manage, and communicate risk and avoid potential catastrophic conditions that can lead to loss of life, property and the environment.

Risk management helps organizations:
1. Identify critical and non-critical risks.
2. Document each risk in-depth.
3. Log all risks and notify management of their severity.
4. Take action to reduce the likelihood of risks occurring.
5. Reduce the impact on business, life, and the environment.

My talk focused on integrated management systems and how they can leverage risk and liability and support sustainability in the business marketplace. The audience was attentive to be sure, and I listened and observed delegates listen to several other fantastic presentations on corporate social responsibility, carbon management and sustainability. My impression however is that while there are pockets of excellence in sustainability focused companies, Nigerian businesses are just beginning to think about sustainability as a value-added aspect of their businesses. Perhaps rightly so, many companies in the oil, agricultural and heavy industry sectors continue (especially the majority small to medium-sized and under-resource companies) are focusing on the basic critical issues of life safety in the workplace, education and meeting basic environmental compliance operations first. To meet this pressing need, organizations have developed world-class frameworks of occupational, health, safety and environmental risk management and despite rampant complaints of lax enforcement of labour and environmental protection laws, the Nigerian government has implemented its corporate governance policies (similar to the U.S Sarbanes-Oxley provisions) that recognize CSR and reporting obligations.

I am firmly of the belief that companies must take care of these basic health and safety issues and lay a firm foundational framework for continual improvement first before they can progress along the sustainability journey. The central themes I heard about how this can be accomplished are through increasing monitoring, education, awareness building, management accountability and trust. Regarding sustainability, it makes little sense force feeding a business approach that has little immediate bearing on managing organizations immediate risks. One must be able to manage the snakes; you know one by one and step by cautious step.

Be patient Nigeria. You have such great resources, professionals hungry to learn, and have fantastic opportunities to excel in the sustainability space in the years ahead. I have been truly blessed and humbled to have been able to participate at this conference and hope to be able to hear of great things coming out of Nigeria in the coming years.

Chapter 23: Viewing Compliance as Opportunity

The first steps companies must take on the long march to sustainability usually arise from the law. Compliance is complicated. Environmental regulations vary by country, by state or region, and even by city. (In 2007 San Francisco banned supermarkets from using plastic bags at checkout; United Kingdom still hasn't.) In addition to legal standards, enterprises feel pressured to abide by voluntary codes; general ones, such as the Greenhouse Gas Protocol, and sector-specific ones, such as the Forest Stewardship Council code and the Electronic Product Environmental Assessment Tool; that non-governmental agencies and industry groups have drawn up over the past two decades. These standards are more stringent than most countries' laws, particularly when they apply to cross-border trade.

It's tempting to adhere to the lowest environmental standards for as long as possible however, it's smarter to comply with the most stringent rules, and to do so before they are enforced. This yields substantial first-mover advantages in terms of fostering innovation. For example, automobile manufacturers in the United States take two or three years to develop a new car model. If GM, Ford, or Chrysler had embraced the California Air Resources Board's fuel consumption and emissions standards when they were first proposed, in 2002, it would be two or three design cycles ahead of its rivals today and poised to pull further ahead by 2016, when those guidelines will become the basis of U.S. law.

Enterprises that focus on meeting emerging norms gain more time to experiment with materials, technologies, and processes. For instance, in the early 1990s Hewlett-Packard realized that because led is toxic, governments would one day ban led solders. Over the following decade it experimented with alternatives, and by 2006 the company had created solders that are an amalgam of tin, silver, and copper, and even developed chemical agents to tackle the problems of oxidization and tarnishing during the soldering process. Thus HP was able to comply with the European Union's Restriction of Hazardous Substances Directive, which regulates the use of led in electronics products, as soon as it took effect, in July 2006.

Contrary to popular perceptions, conforming to the gold standard globally actually saves companies money. When they comply with the least stringent standards, enterprises must manage component sourcing, production, and logistics separately for each market, because rules differ by country. HP, Cisco, and other companies that enforce a single norm at all their manufacturing facilities worldwide benefit from economies of scale and can optimize supply chain operations. The common norm must logically be the toughest.

A Few Simple Rules

Smart corporations follow these simple rules in their effort to become sustainable.

Don't start from the present. If the starting point is the current approach to business, the view of the future is likely to be an optimistic extrapolation. It's better to start from the future. Once senior managers establish a consensus about the shape of things to come, they can fold that future into the present. They should ask. What are the milestones on the path to our desired future? What steps can we take today that will enable us to get there? How will we know that we are moving in that direction?

Ensure that learning precedes investments. Top management's interest in sustainability sometimes leads to investments in projects without an understanding of how to execute them. Smart companies start small, learn fast, and scale rapidly. Each step is broken into three phases; experiments and pilots, debriefing and learning, and scaling. These companies benchmark, but the goal is to develop next practices not merely mimic best practices.

Stay wedded to the goal while constantly adjusting tactics. Smart executives accept that they will have to make many tactical adjustments along the way. A journey that takes companies through five stages and lasts a decade or more can't be completed without course corrections and major changes. Although directional consistency is important, tactical flexibility is critical.

Build collaborative capacity. Few innovations, be they to comply with regulations or to create a new line of products, can be developed in today's world unless companies form alliances with other businesses, nongovernmental organizations, and governments. Success often depends on executives' ability to create new mechanisms for developing products, distributing them, and sharing revenues.

Use a global presence to experiment. Multinational corporations enjoy an advantage in that they can experiment overseas as well as at home. The governments of many developing countries have become concerned about the environment and are encouraging companies to introduce sustainable products and processes, especially for those at the bottom of the pyramid. It's easier for global enterprises to foster innovation in emerging markets, where there are fewer entrenched systems or traditional mind-sets to overcome.

Companies can turn antagonistic regulators into allies by leading the way. For instance, HP has helped shape many environmental regulations in Europe, and it uses the resulting brownie points to its advantage when necessary. In 2001 the European Union told hardware manufacturers that after January 2006 they could not use hexavalent chromium which increases the risk of cancer in anyone who comes in contact with it as an anticorrosion coating. Like its rivals, HP felt that the industry needed more time to develop an alternative. The company was able to persuade regulators to postpone the ban by one year so that it could complete trials on organic and trivalent chromium coatings. This saved it money, and HP used the time to transfer the technology to more than one vendor. The vendors competed to supply the new coatings, which helped reduce HP's costs.

Companies in the vanguard of compliance naturally spot business opportunities first. In 2002 HP learned that Europe's Waste Electrical and Electronic Equipment regulations would require hardware manufacturers to pay for the cost of recycling products in proportion to their sales. Calculating that the government-sponsored recycling arrangements were going to be expensive, HP teamed up with three electronics makers; Sony, Braun, and Electrolux to create the private European Recycling Platform. In 2007 the platform, which works with more than 1,000 companies in 30 countries, recycled about 20% of the equipment covered by the WEEE Directive. Partly because of the scale of its operations, the platform's charges are about 55% lower than those of its rivals. Not only did HP save more than $100 million from 2003 to 2007, but it enhanced its reputation with consumers, policy makers, and the electronics industry by coming up with the idea.

Chapter 24: Sustainability Return on Investment

Today's corporations and institutions are increasingly responsible for the environmental and societal impacts of their decisions. That is why you need a decision-making methodology that incorporates broad social, economic and environmental forecasting; one that helps you minimize cost and maximize return on investment (ROI).

Sustainability Return on Investment (S-ROI) integrates the ability to measure the social, economic and environmental returns on sustainability initiatives. The methodology is designed to examine a decision-making process from the viewpoint of multiple expert stakeholders, while maximizing return for as many of them as possible.

The methodology allows for the enumeration of uncertain events with their concurrent costs and benefits. The decision-maker obtains a financial picture of the future of a decision that includes best case, worst case, and most probable ranges of return on investment. This methodology deals with both internal costs (those borne by the company) and external costs (those borne by society), allowing decision-makers to use both aspects as appropriate.

The output gives key decision makers across functions the actionable information they need to justify various forms of investment in socially and environmentally responsible activity, while avoiding "burden shifting," where one harmful impact is simply traded for another.

In many cases, decision-makers use the S-ROI methodology to determine whether there is a business justification for initiatives that don't show a positive ROI based on traditional costing methods.

Here are a few examples:

1. The S-ROI methodology showed positive ROI for the Japanese government to support construction of a biogas factory.

2. The S-ROI methodology showed positive ROI for a mining company in South Africa, where investing in HIV/AIDS education didn't make sense from a traditional accounting perspective. When costs to society were included in the ROI calculation, the project was clearly net positive in its result, both for the company and for the surrounding communities.

S-ROI is also helpful when the decision may be affected by uncertainty; fluctuations in fuel prices, potential new regulations, changes in fads, etc. Other example decisions where S-ROI analysis can help ensure maximum benefits include whether to invest in pollution prevention devices, whether to make an acquisition, or even whether or not to expand a facility.

Chapter 25: Making Value Chains Sustainable

Once companies have learned to keep pace with regulation, they become more proactive about environmental issues. Many then focus on reducing the consumption of non-renewable resources such as coal, petroleum, and natural gas along with renewable resources such as water and timber. The drive to be more efficient extends from manufacturing facilities and offices to the value chain. At this stage, corporations work with suppliers and retailers to develop eco-friendly raw materials and components and reduce waste. The initial aim is usually to create a better image, but most corporations end up reducing costs or creating new businesses as well. That is particularly helpful in difficult economic times, when corporations are desperate to boost profits.

Companies develop sustainable operations by analyzing each link in the value chain. First they make changes in obvious areas, such as supply chains, and then they move to less obvious suspects, such as returned products.

Supply chains

Most large corporations induce suppliers to become environment-conscious by offering them incentives. For instance, responding to people's concerns about the destruction of rain forests and wetlands, multinational corporations such as Cargill and

Unilever have invested in technology development and worked with farmers to develop sustainable practices in the cultivation of palm oil, soybeans, and other agricultural commodities. This has resulted in techniques to improve crop yields and seed production.

Some companies in the West have also started laying down the law. For example, in October 2008 Lee Scott, then Wal-Mart's CEO, gave more than 1,000 suppliers in China a directive. Reduce waste and emissions; cut packaging costs by 5% by 2013; and increase the energy efficiency of products supplied to Wal-Mart stores by 25% in three years' time. In like vein, Unilever has declared that by 2015 it will be purchasing palm oil and tea only from sustainable sources, and Staples intends that most of its paper-based products will come from sustainable-yield forests by 2015.

Tools such as enterprise carbon management, carbon and energy footprint analysis, and life-cycle assessment help companies identify the sources of waste in supply chains. Life-cycle assessment is particularly useful. It captures the environment-related inputs and outputs of entire value chains, from raw-materials supply through product use to returns. This has helped companies discover, for instance, that vendors consume as much as 80% of the energy, water, and other resources used by a supply chain, and that they must be a priority in the drive to create sustainable operations.

Operations

Central to building a sustainable supply chain are operational innovations that lead to greater energy efficiency and reduce companies' dependence on fossil fuels. Take the case of FedEx, which deploys a fleet of 700 aircraft and 44,000 motorized vehicles that consume 4 million gallons of fuel a day. Despite the global slowdown, the company is replacing old aircraft with Boeing 757s as part of its Fuel Sense program. This will reduce the company's fuel consumption by 36% while increasing capacity by 20%. It is also introducing Boeing 777s, which will reduce fuel consumption by a further 18%. FedEx has developed a set of 30 software programs that help optimize aircraft schedules, flight routes, the amount of extra fuel on board, and so on. The company has set up 1.5-megawatt solar-energy systems at its distribution hubs in California and Cologne, Germany. It uses hybrid vans that are 42% more fuel efficient than conventional trucks and has replaced more than 25% of its fleet with smaller, more fuel-efficient vehicles. Following some other pioneers, FedEx recently turned its energy-saving expertise into a stand-alone consulting business that, it hopes, will become a profit centre.

Vendors consume as much as 80% of the resources used by a supply chain.

Workplaces

Partly because of environmental concerns, some corporations encourage employees to work from home. This leads to reductions in travel time, travel costs, and energy use. One-tenth of the corporations in our sample had from 21% to 50% of their employees telecommuting regularly. Of IBM's 320,000 employees, 25% telecommute, which leads to an annual savings of $700 million in real estate costs alone. AT&T estimates that it saves $550 million annually as a result of telecommuting. Productivity rises by 10% to 20%, and job satisfaction also increases when people telecommute up to three days a week. For example, at the health-care services provider McKesson, the group that reported the highest job satisfaction in 2007 consisted of 1,000 nurses who worked from home.

Returns

Concerns about cutting waste invariably spark companies' interest in product returns. In the United States, returns reduce corporate profitability by an average of about 4% a year. Instead of scrapping returned products, companies at this stage try to recapture some of the lost value by reusing them. Not only can this turn a cost centre into a profitable business, but the change in attitude signals that the company is more concerned about preventing environmental damage and reducing waste than it is about cannibalizing sales.

Cisco, for example, had traditionally regarded the used equipment it received as scrap and recycled it at a cost of about $8 million a year. Four years ago it tried to find uses for the equipment, mainly because 80% of the returns were in working condition. A value-recovery team at Cisco identified internal customers that included its customer service organization, which supports warranty claims and service contracts, and the labs that provide technical support, training, and product demonstrations. In 2005 Cisco designated the recycling group as a business unit, set clear objectives for it, and drew up a notional P&L account. As a result, the reuse of equipment rose from 5% in 2004 to 45% in 2008, and Cisco's recycling costs fell by 40%. The unit has become a profit centre that contributed $100 million to Cisco's bottom line in 2008.

When they create environment-friendly value chains, companies uncover the monetary benefits that energy efficiency and waste reduction can deliver. They also learn to build mechanisms that link sustainability initiatives to business results, as the Cisco example shows. As a result, environmental concerns take root within business units, allowing executives to tackle the next big challenge.

Chapter 26: Creating Next Practice Platforms

Next practices change existing paradigms. To develop innovations that lead to next practices, executives must question the implicit assumptions behind current practices. This is exactly what led to today's industrial and services economy. Somebody once asked. Can we create a carriage that moves without horses pulling it? Can we fly like birds? Can we dive like whales? By questioning the status quo, people and companies have changed it. In like vein, we must ask questions about scarce resources. Can we develop waterless detergents? Can we breed rice that grows without water? Can biodegradable packaging help seed the earth with plants and trees?

Sustainability can lead to interesting next-practice platforms. One is emerging at the intersection of the internet and energy management. Called the smart grid, it uses digital technology to manage power generation, transmission, and distribution from all types of sources along with consumer demand. The smart grid will lead to lower costs as well as the more efficient use of energy. The concept has been around for years, but the huge investments going into it today will soon make it a reality. The grid will allow companies to optimize the energy use of computers, network devices, machinery, telephones, and building equipment, through meters, sensors, and applications. It will also enable the development of cross-industry platforms to manage the energy needs of cities, companies, buildings, and households. Technology vendors such as Cisco, HP, Dell, and IBM are already investing to develop these platforms, as are utilities like Duke Energy, SoCal Edison, and Florida Power & Light etc.

Two enterprise-wide initiatives help companies become sustainable.

1. When a company's top management team decides to focus on the problem, change happens quickly. For instance, in 2005 General Electric's CEO, Jeff Immelt, declared that the company would focus on tackling environmental issues. Since then every GE business has tried to move up the sustainability ladder, which has helped the conglomerate take the lead in several industries.

2. Recruiting and retaining the right kind of people is important. Recent research suggests that three-fourths of workforce entrants in the United States regard social responsibility and environmental commitment as important criteria in selecting employers. People who are happy about their employers' positions on those issues also enjoy working for them. Thus companies that try to become sustainable may well find it easier to hire and retain talent.

Leadership and talent are critical for developing a low-carbon economy. The current economic system has placed enormous pressure on the planet while catering to the needs of only about a quarter of the people on it, but over the next decade twice that number will become consumers and producers.

Traditional approaches to business will collapse, and companies will have to develop innovative solutions. That will happen only when executives recognize a simple truth. Sustainability equals Innovation.

Chapter 27: Navigating Sustainable Supply Chain Management in China

2014 marked a watershed moment in supply chain sourcing among worldwide manufacturers and retailers. Sustainability observers and practitioners read nearly weekly announcements of yet another major manufacturer or retailer setting the bar for greener supply chain management. With a much greater focus on monitoring, measurement and verification, retailers and manufacturers Wal-Mart, Marks and Spencer, IBM, Proctor and Gamble, Kaiser Permanente, Puma, Ford, Intel, Pepsi, Kimberly-Clark, Unilever, Johnson & Johnson among many others made major announcements concerning efforts to engage, collaborate and track supplier/vendor sustainability efforts, especially those involving overseas operations. Central to each of these organizations is how suppliers and vendors impact the large companies' carbon footprint, in addition to other major value chain concerns such as material and water resource use, waste management and labour/human rights issues. Meanwhile, efforts from China's manufacturing sector regarding sustainable sourcing and procurement, was at best, mixed with regard to proactive sustainability. From my perspective as a sustainability practitioner (with a passion in supply chain management), the challenges that foreign businesses with manufacturing relationships in China can be daunting. Recent events concerning Apple Computers alleged lax supplier oversight and reported supplier human rights and environmental violations only shows a microcosm of the depth of the challenges that suppliers face in managing or influencing these issues on the ground. Apple did the right thing by transparently releasing its Apple Supplier Responsibility 2011 Progress Report, which underscored just how challenging and difficult multi-tiered supply chain management can be. But all is certainly not lost and many companies have in recent years begun to navigate the green supply chain waters in China.

According to a World Resources Institute (WRI) White Paper issued in the fall of 2010, China faces a number of supply chain challenges. First, the recent spate of reports alleging employee labour and environmental violations can place manufacturing partnerships with global corporations at risk. According to the report, Chinese suppliers that are unable to meet the environmental performance standards of green supply chain companies may not be able to continue to do business with such firms. Wal-Mart has already gone on record, announcing that it will no longer purchase from Chinese suppliers with poor environmental performance records. In order to be a supplier to Wal-Mart, Chinese companies must now provide certification of their compliance with China's environmental laws and regulations.

Wal-Mart, like many other IT and apparel manufacturers also conducts audits on a factory's performance against specific environmental and sustainability performance criteria, such as air emissions, water discharge, management of toxic substances and hazardous waste disposal. These actions are extremely significant as Wal-Mart procures from over 10,000 Chinese suppliers. This increased scrutiny on environmental and corporate social responsibility through supplier scoring and sustainability indexing, says the WRI report may trump price, quality, and delivery time as a decisive factor in a supplier's success in winning a purchasing contract.

Chinese Government Stepping Up Enforcement

The WRI report indicated that the Chinese State Council is directing key government agencies, including the National Development and Reform Commission, the Ministry of Finance, and the Ministry of Environmental Protection to prohibit tax incentives, restrict exports and raise fees for energy intensive and polluting industries, such as steel, cement, and minerals extraction. Also, it's been reported in the past years that the People's Bank of China and the Ministry of Environmental Protection are also working with local Chinese banks to implement the 'Green Credit' program, which prevents loans to Chinese firms with poor environmental performance records. In addition, the National Development and Reform Commission and the Ministry of Finance have issued a notice to all Chinese central and local governments to purchase goods from suppliers that are 'energy efficient'. Finally, on a local level, governments have developed preferred supplier lists for companies producing environmental-friendly products for their purchasing needs.

Supplier Challenges Are Not Just Environmental

A China Supply Chain Council survey conducted in 2009 identified a huge gap in knowledge between (1) clear understanding of which environmental issues posed the greatest risk (2) what to do to manage significant environmental risks. Also, nearly 40% of the company's surveyed thought sustainability to be cost prohibitive, too complicated or where particular expertise was lacking don't have the expertise (on the other hand 60% did!). Two-thirds of respondents did consider sustainability to be a supply chain priority, although many were not confident of the return on investment however, more than half of the respondents reported that they had begun collaborating with their larger supply chain partners. In fact, according to the World Resources Institute White Paper, despite increasing pressures to improve their environmental performance, Chinese suppliers face many financial challenges to operating in a more sustainable manner

World Resources Institute White paper notes increasing non-environmental pressures, including.

1. Extended green investment payback: While improving resource consumption, such as energy and water, provides long-term cost savings, the payback for making such environmental investments may be as long as three years, which is financially impossible for many Chinese suppliers.

2. Lack of financial incentives from green supply chain buyers: Multinational buyers are often unwilling to change purchasing commitments and long-term purchasing contracts to Chinese suppliers that make the investments to improve their environmental performance.

3. Rising operational costs: Chinese suppliers face rising resource and labour costs. For example, factory wages have increased at an average annual rate of 25 percent during 2007 to 2013. Rising costs dissuade suppliers from making environmental investments which may raise operating costs.

4. Limited access to finance: The majority of Chinese suppliers are small and medium-scale enterprises (SMEs) with limited access to formal financing channels such as bank loans. Chinese SMEs account for less than 10 percent of all bank lending in China, and as a result, Chinese suppliers frequently do not have the capital to make the necessary environmental investments.

5. Intense domestic and global competition: Chinese suppliers face intense competition from thousands of firms, both domestic and international, within their industries. This intense competition puts constant pressure on suppliers to cut costs, which can include environmental protections, in an effort to stay in business.

Leveraging the Supply Chain to Gain "Reciprocal Value"

Leading edge, sustainability minded and innovative companies have found "reciprocal value" through enhanced product differentiation, reputation management and customer loyalty. I highlighted in earlier chapter of this book the model efforts that GE has implemented with its China based suppliers to implant responsible and environmentally proactive manufacturing into their operations. GE's comprehensive supplier assessment program evaluates suppliers in China and other developing economies for environment, health and safety, labour, security and human rights issues. GE has leaned on its thousands of suppliers to obtain the appropriate environmental and labour permits, improve their environmental compliance and overall performance. In addition, GE and other multi-national companies (including Wal-Mart, Honeywell, Citibank and SABIC Innovative Plastics) have partnered to create the EHS Academy in Guangdong province. The objective of this no-profit venture is to create a more well-trained and capable workforce of environmental, health and safety professionals.

In this book we highlighted the critical needs for increased supply chain collaboration among the world's largest manufacturers in order to effectively operationalize sustainability in Chinese manufacturing plants. This is especially evident for large worldwide manufacturers operating subcontractor arrangements in developing nations and "tiger economies", such as India, Mexico and China (and the rest of Southeast

Asia). Global manufacturer efforts underscore how successful greening efforts in supply chains can be based on value creation through the sharing of intelligence and know-how about environmental and emerging regulatory issues and emerging technologies.

Suppliers and customers stand so much to gain from collaboratively strengthening each other's performance and sharing cost of ownership and social license to operate. But as I have stated before, supply chain sustainability and corporate governance must first be driven by the originating product designers and manufacturers that rely on deep tiers of suppliers and vendors in far-away places for their products.

Chapter 28: Sustainable Supply Chain Creates a Competitive Advantage Worldwide

Consumer awareness is growing around the world and, combined with the financial benefits, a sustainable supply chain makes long term business sense

Companies must confront the reality that their supply chains can no longer be opaque. Stakeholders demand more accountability and want to know about suppliers' ethics pertaining to workers and the environment. A cleaner and more responsible supply chain, however, is not just about satisfying consumer concerns in wealthier developed countries. As the middle class grows in emerging markets, consumer awareness about sustainability is on the rise; in fact, such expectations in developing countries are often higher than in established economies.

Add the financial benefits of energy and resource efficiencies, and supply chain sustainability creates a competitive advantage for companies worldwide. Yet as a company's supply chain team stamps out waste and inefficiencies, communication with suppliers is key to having all stakeholders on board. Rather than a top-down approach with a list of demands, companies have got to advise, counsel and even support their suppliers, and their communities, logistically and sometimes even financially.

The home furnishings giant IKEA is one company that works with suppliers on a variety of challenges, from energy efficiency to sourcing materials responsibly. During a conversation one of my colleagues had with IKEA's chief sustainability officer recently, he explained how much of the company's sustainable supply chain work is underway in countries where much of its supplier base is located.

For example, IKEA has pledged to invest €1.5b in renewable energy technologies. Much of that investment is underway in China, where IKEA announced it would install solar panels on all company-owned buildings throughout the country. But IKEA is also expanding the programme to its supplier base. With the cost of solar more competitive than ever before, IKEA's professionals are reaching out to Chinese suppliers to find the best systems with the quickest returns on investment. The results not only lower the carbon footprint of IKEA's supply chain, but save the company and suppliers money from reduced utility costs.

Natural resource depletion is a long term challenge to IKEA's business model and IKEA claimed that they are fine-tuning the sourcing of wood for its products. The company has hired forestry specialists who work with suppliers on educating them about more responsible wood procurement practices. IKEA suppliers in turn must report the origin of their wood every four months and then are subjected to audits to which they have only 48 hours to report the origins of their wood. Meanwhile IKEA conducts wider supply chain audits so that the company can trace the origin of wood all the way back to the actual forest. In addition to wood, IKEA also trains suppliers and other stakeholders on issues related to waste, energy and water.

Water has emerged as a huge challenge for supply chains. PepsiCo is one company taking the lead on working with suppliers to confront water scarcity. All of its products from crisps to soda rely on this most precious resource, and as the company expands into markets, its new operations end up in water-stressed regions. To that end, PepsiCo's sustainability team reaches out across its value chain, including suppliers, to find new ways for more efficient water use; and the company works with charities to ensure the communities in which it operates have adequate access to fresh water.

PepsiCo extends advice to its supply chain's furthest reaches. The company partners with NGOs to develop crops most appropriate for local climates that end up in the company's products. In Mexico, PepsiCo has supported over 800 small farmers with micro loans and other funding schemes to grow sunflowers for oil used to make food products marketed in the Latin American market. Throughout Ethiopia, the company works with USAID and the World Food Programme to cultivate chickpeas, a crop that requires minimal water while self-fertilising the soil with minimal reliance on nitrates or phosphates. A focus on Africa only makes sense for PepsiCo, which has plans to expand into that market and sell products to consumers of all income levels.

The other half of Pepsi's two-pronged approach is working with NGOs to ensure stable water supplies. In India, Water.org provides micro loans, funded by PepsiCo's philanthropic arm, to families within and beyond communities in which it operates. Other companies could take a page from PepsiCo's approach, which builds trust with stakeholders by working with the company's most challenged suppliers as well as demonstrating to communities the company is an engaged partner, not an imperious foreign business extracting resources to other markets.

After globalisation of markets, organisations are striving very hard to improve their competitiveness in a globalised market. In the recent past, Supply Chain Management (SCM) has played very crucial role in improving efficiency of different operations across whole value chain. Most of the studies on SCM have focused on diverse topics including inventory control, risk management, sustainable SCM, supply chain network etc. In present economic scenario, organisations are trying to achieve sustainable competitiveness in global markets. Sustainability incorporates the concepts of economic, social, and environmental performance. Green Supply Chain Management (GSCM) practices comprise green design, reducing energy consumption, reusing/recycling material and packaging, reverse logistics and environmental collaboration in the supply chain. To achieve sustainable competitiveness, many researchers have advocated for Green Supply Chain Management.

In addition to it, many countries have made stringent environmental regulations for encouraging green products and practices. For example, some countries require a certificate for wood products in order to show that their harvest does not harm their forest's sustainable development. Another example is that shoes made in Fujian, a province in southeast China, could not be exported because the glue used in shoe manufacturing does not satisfy the environmental requirements of the customers. Some countries, including Japan, United States, Netherlands, Norway, France and Sweden, have also put forward different environmental requirements for the fabrics and dyes of clothes imported from China. It is therefore a great interest to explore how locational context enables firms to raise their sustainable competitiveness through Green Supply Management and/or how firms' behaviour including Green Supply Management Strategies raises a locations' sustainable competitiveness.

Chapter 29: German Sustainable Development Strategy

In fact, the German government has made a formal commitment towards sustainable supply chain management (SSCM) through the German Sustainable Development Strategy (GSDS).

The purpose of the GSDS was to classify procurement activities and find a means of attaching measurable aspects to them. The five procurement activities that are accountable within the GSDS are.
1. Reducing logistics and freight intensity.
2. Reduce land use.
3. Implications against partners because of sustainable aspects.
4. Quality of employee working conditions.
5. Enhancing quality of employment.

Within reducing logistics and freight intensity, the GSDS measures greenhouse gas emissions, use and disposal of truck tires, and measuring driver behaviour according to environmentally conscious driving standards.

The purpose of measuring land use falls on the dimensions of reducing energy waste, increasing renewable energy for warehouses and factories, and actively considering the environment in property choices.

Under the GSDS, sustainable partner activities used for measurement include whether they use rail or ship cargo, whether partners use environmentally friendly transport services which does fall into the first activity above, and whether partners use combined transport to reduce freight intensity.

For a long time, the procurement industry was overrun with allegations of poor working conditions and low access to essential services however through the GSDS, German firms are observing whether employees are given services during the day and on weekends. German firms are also attempting to maintain a standard level of pay and minimize the use of temporary workers and the final activity dimension measurable under the GSDS involves education and enhancing qualified employment in all levels of supply chain operations.

The GSDS is indeed a highly ambitious effort in attempting to measure sustainable practices within the German procurement industry.

Sustainable procurement is incredibly important for the future success of an organizations supply chain, but also for our environment and planet. The difficulties are not unheard of when considering green procurement strategies, however the strategic advantage that one could attain can last well into the future.

Chapter 30: Developing a National Sustainable Good Greenhouse Gas Inventory

Good greenhouse gas (GHG) inventories are crucial; by knowing the amounts and sources of GHGs, policymakers can prioritize and design domestic strategies to reduce them. That is why many countries are working to develop a national inventory system that is robust and sustainable. Many developing ("non-Annex I") countries, in particular, are seeking to improve their national inventory systems in an effort to support domestic low-carbon goals, as well as start submitting regular reports to the UNFCCC that include a national GHG inventory.

Although there is no "one-size-fits-all" solution to developing a sustainable national GHG inventory system, countries can learn from each other's experiences. What worked and why? What hasn't worked and why? And how have countries built their capabilities for compiling a national inventory over time?

Seven Emerging Good Practices

The seven emerging good practices for developing sustainable national GHG inventories are:

1. Sustain institutional arrangements. Having the same institutions involved in the national inventory process each time can help build a strong foundation for

retaining institutional memory and facilitating oversight and accountability.

2. Identify and empower a lead agency to manage the national GHG inventory process. It's not enough for a national government to assign an agency responsibility for the national inventory; that agency needs to be empowered with clear authority, technical expertise, and sufficient financial resources.

3. Create coordinating institutions for each sector, with well-defined roles, responsibilities, and processes. The lead agency may use sector coordinators or working groups to delegate to other entities some responsibility for portions of the national GHG inventory where technical expertise is more readily available. More importantly, coordination can also result in greater buy-in for the national inventory process, including data collection, promote interagency collaboration, and improve the overall quality of inventory information.

4. Establish detailed institutional mandates and data-sharing agreements that include work schedules. Institutional mandates, agreements, or memoranda of understanding keep everyone from government agencies to external groups on the same page about what they are expected to contribute to the inventory. Setting clear expectations early on could also help save time and money.

5. Archive inventory information and retain institutional memory. Establishing archiving systems should be a first-order priority for countries looking

to develop a sustainable national inventory system. Even if the solution isn't highly technical, capturing the data and methods used, as well as information regarding processes, participants, and lessons learned will make the next inventory process easier.

6. Allocate sufficient, well managed, and sustained financial resources. The strategic allocation of money very much has a role to play in the adoption of almost all the other good practices we identified. One takeaway from our study is that countries often have to leverage other sources of funds, including "in kind" contributions from other federal initiatives, in addition to the money a country may receive from the Global Environment Facility; a funding mechanism for completing UNFCCC reports.

7. Improve the national GHG inventory system over time. This last good practice emphasizes that governments don't need significant resources to initiate a sustainable national inventory system. No national inventory system is perfect and developing a robust inventory system takes time. Countries should start with the resources they have, build out the system in a strategic way, and regularly review to determine what improvements can be made.

Measure to Manage and then Share

When building capacity for national inventories, the country-specific context matters. For example, some countries might have institutions and individuals with GHG measurement expertise whereas other countries do not but have well-established data-sharing arrangements among government agencies; it's important to acknowledge strengths and weaknesses and then build on them.

Understanding the range of national inventory system needs and the potential solutions has value to both countries looking to develop a national inventory system and the capacity-building organizations and donors looking to support them.

Chapter 31: Hope for Sustainable Economies

Consumers have unprecedented opportunity to be active shapers of the products and services they buy and use, rather than passive receivers, taking whatever companies provide. Apples most recent litmus test on corporate social responsibility with its key Chinese supply chain manufacturing partner, Foxconn, and resulting consumer outcry is but just one example of the power that consumers have to sway products manufacturers to alter their business patterns.

At the World Economic Forum (WEF) in Davos, Aron Cramer (CEO and President of Business for Social Responsibility or BSR) observed at one workshop the "fast-changing relationship between businesses and consumers. The question on the minds of many of the business executives in the room was "is this good or bad for business".

The answer to this particular either/or question is undoubtedly both. Companies that stay ahead of this curve by involving consumers in product design; providing transparent information about the social and environmental content of these products, and looking at new models to provide value in new ways will prosper. Those that don't will find growth hard to come by."

Scaling Consumption in a Smart and Sustainable Way

The WEF has devoted a great deal of attention to the issue of scaling consumption sustainably as the world economy shifts both demographically and economically. WEF examines these issues in a report entitled; "More with Less: Scaling Sustainable Consumption and Resource Efficiency". The study properly takes a "systems view" of sustainable consumption. In other words, rather than focusing just on the demand side, WEF looks at the challenges and possible solutions through a value-chain centric lens of what they describe as
1. Consumer engagement (demand).
2. Value chains and upstream action (supply).
3. Policies and an enabling environment to accelerate change (rules of the game).

Making your business sustainable in today's world is an absolute imperative. The business case for sustainable growth is clearer than ever and the urgency of the issues we face means that business leaders have no choice but to act.

As WEF explains, The main outcome is the identification of key focus areas for business leadership through concrete goals and collaboration across industries. For this report, WEF engaged with chief executive officers, business leaders and experts worldwide, seeking answers and thoughts centred six key questions.
1. What are the key trends in sustainable consumption?

2. What is the size of the opportunity for countries, companies and consumers?
3. What are the barriers to scaling existing models of sustainable consumption?
4. What does getting to scale look like?
5. What new solutions are needed to get to scale in sustainable consumption?
6. How can we achieve scale by working collectively and creating action on new fronts?

Barriers, Mind Sets and Complexities- Oh My!

To no surprise, the report identified a number of internal and external barriers to staving and influencing scalable and sustainable consumption, notably (according to the report).

1. Consumers lack incentives for sustainable consumption and are confused by mixed messages. The study noted that one survey of British consumers indicated that 70% were uncertain about the environmental performance of the products they buy. I have seen similar surveys in the United States that compare with the British results

2. Supply chains are complex, opaque and interconnected. Deep supply chains, like Apples or the textile industry, create many complexities that place limits to in certainties sustainable sourcing.

3. Technology remains costly and inadequately deployed. The study notes that "Fewer than 20 facilities in the world are certified to melt down and recycle the cathode ray tubes of old television sets,

and all are in Asia. E-waste, which at present largely originates in the US and Europe will travel across multiple countries and continents for recycling; putting the environmental benefits into question and causing additional social concerns". That being said, more collaborative enterprises across industries and economies can replace the linear economies that characterize western industrial nations, and create more opportunities to expand technologies further and wider.

4. Policy incentives remain weak. The report notes that "trade systems and tariffs rarely differentiate between unsustainable and more sustainable alternatives, preventing a potential increase of 7-13% in the traded volumes of sustainable products.

5. Short-termism dominates the landscape, and traction in fast-growing markets remains low. Typical of capitalism and free enterprise, most companies growth targets rarely look out further than a few years, and seek short term gains to keep shareholders happy. The WEF report noted that "55% of FTSE 100 company sustainability targets were to be achieved within 1-2 year timeframes, while only 18% looked out to 2018–2020".

Solutions for Scaling Economies (Source, WEF, 2012)

Moving Toward a Circular Economy

Something else also happened; on the way to the Forum; well actually at the Forum that may offer some insights and solutions that are discussed in the WEF report. At Davos, Ellen MacArthur, head of the non-profit Ellen MacArthur Foundation, suggested that while rapid technological evolution across all major industry sectors, was taking place very little change within the economic model itself has been occurring. The economy is still based on a linear take, make and dispose model. A new report Towards the Circular Economy, analyzes the international business case behind the idea of shifting from a linear to a more circular economy.

The essence of the circular economy lies in designing goods using technical materials to facilitate disassembly and re-use, and structuring business models so manufacturers can reap rewards from collecting and refurbishing, remanufacturing, or redistributing products they make. In this model all things are made to be made again, ultimately using energy from renewable sources and in a less toxic manner. Companies shift to focusing on selling performance in the place of product, and consumers now become users.

The time for sustainable consumption is now. The need to develop new consumption patterns is the mother of all innovation challenges. The race to dematerialize is on. Some of this will come from the digital revolution, as newspapers can now be delivered wirelessly to e-readers instead of plopping dead trees on the doorstep. But some of the innovation will come from redesigning business models.

Are you, as consumer, as manufacturer, product designer or corporate executive, or even as fellow Planet-eer, ready to help make that change? We can change the rules of the game together, for a stronger, more circular economy. As Captain Planet says, "The Power is Yours".

Chapter 32: Green Supply Chain Management Requires Less Procrastination

Admit it, we have all done it! Procrastinated. Waited until the brink of a bad outcome. Not taken the time to thoughtfully, proactively, pragmatically complete an assignment, implement a new 'leading edge' technology or launch a disruptively innovative initiative. Instead we react, overlook great ideas for something less, produce a less articulate response to an inquiry, or implement a semi thought out idea.

Even in the business world, whether in supply chain management or in adoption of the 'triple bottom line' in business strategy, there are leaders and there are laggards. Innovators and adopters. I was reminded of this when I ran across a research paper that was published in "Sustainability" Journal recently. The article, "Supply Chain Management and Sustainability: Procrastinating Integration in Mainstream Research" presents the results of a study conducted by several university researchers in The Netherlands. The researchers noted that "procrastination can be viewed as the result of several processes, determined not only by individual personality, but also by the following factors:
1. Availability of information.
2. Availability of opportunities and resources.
3. Skills and abilities.
4. Dependence on cooperation with others."

In addition, in a review of more than 100 additional studies on procrastination, the following additional items were found to likely to influence procrastination:
1. The nature of the task.
2. The context of the issue.

It is these last two issues that the authors raised as primary reasons for procrastination, especially regarding embedding sustainability research and practices in supply chain operations and management. The authors found that "the nature of the task", because it's often complex and requires many internal and external stakeholders, and therefore tends to "generate conflicts". Also, the roots of supply chain management and related research are generally grounded in operations management and operations/logistics. Therefore, the researchers noted that environmental and social aspects of supply chain management are foreign, "out of context" and not wholly integrated into supply chain management and research. I would also argue that dependence on others is a key issue as well given the widespread, outward facing challenges associated with supply chain coordination.

So what this means is that if a concept is foreign or unfamiliar or "out of context" it's either set aside as being non-value added. Also because of some of the complexities often inherent in grasping and applying sustainability concepts, some just throw up their hands and say "I have no time for this". This in turn can lead to procrastination in the real-world application of sustainability in supply chain management.

In a study conducted during the height of the recession (late 2009), GTM Research found that despite its growing prominence, "sustainability is not a core part of most companies' strategies or …a prime driver of their supply chain agendas." The study found that sustainability lies in the middle of the pack of supply chain priorities, behind cost cutting. The graphic presents a "leaders vs. laggards" scenario. The 23% difference between leaders and laggards related to sustainability initiative implementation is large and underscores the work that remains to advance the "value proposition" for sustainability in supply chain management.

In this book we described positive aspects of adopting whole systems-based, collaborative and transparent approaches to sustainable sourcing and manufacturing, and green logistics. Sustainable thinking in supply chain management also value chain practices supports environmental and social responsibility; so why aren't more companies adopting these methods?

I know how many of the leaders are implementing greener and more sustainable supply chain practices in their respective markets and we have mentioned them in this book. Laggards? Well you know who you are, but I am not pointing fingers.

While the future looks bright for a "greener" perspective in supply chain management, there still

remains a stigma that a sustainable value chain is a costly one. In reality, there may be some up-front costs associated with some initiatives; very true but companies must take a longer view and pencil out the ROI of supply chain sustainability best practices and it's possible by taking a leap and reaping the benefits. I am confident that those organizations who wish to lead (and stop procrastinating!) will find a great many benefits including.
1. Less resource intensive product designs.
2. Better supply chain planning and network optimization.
3. Better coordinated warehousing and distribution.
4. More advanced and innovative reverse logistics options.

Those who choose to lead will realize significant cost savings, improved efficiencies and a more secure and profitable future.

Chapter 33: Triple Bottom Line

We cannot conclude this book without defining and explaining what the Triple Bottom Line (TBL) is and how it work.

The TBL is an accounting framework that incorporates three dimensions of performance; social, environmental and financial. This differs from traditional reporting frameworks as it includes ecological (or environmental) and social measures that can be difficult to assign appropriate means of measurement. The TBL dimensions are also commonly called the three Ps: people, planet and profits. We will refer to these as the 3Ps.

Well before Elkington introduced the sustainability concept as "triple bottom line," environmentalists wrestled with measures of, and frameworks for, sustainability. Academic disciplines organized around sustainability have multiplied over the last 30 years. People inside and outside academia who have studied and practiced sustainability would agree with the general definition of Andrew Savitz for TBL. The TBL "captures the essence of sustainability by measuring the impact of an organization's activities on the world ... including both its profitability and shareholder values and its social, human and environmental capital.

The trick isn't defining TBL. The trick is measuring it.

Calculating the TBL

The 3Ps do not have a common unit of measure. Profits are measured in dollars. What is social capital measured in? What about environmental or ecological health? Finding a common unit of measurement is a challenge.

Some advocate monetizing all the dimensions of the TBL, including social welfare or environmental damage. While that would have the benefit of having a common unit/dollars many object to putting a dollar value on wetlands or endangered species on strictly philosophical grounds. Others question the method of finding the right price for lost wetlands or endangered species.

Another solution would be to calculate the TBL in terms of an index. In this way, one eliminates the incompatible units issue and, as long as there is a universally accepted accounting method, allows for comparisons between entities, e.g., comparing performance between companies, cities, development projects or some other benchmark.

An example of an index that compares a county versus the nation's performance for a variety of components is the Indiana Business Research Centre's Innovation Index. There remains some subjectivity even when using an index however. For example, how are the index components weighted? Would each "P" get equal weighting? What about the sub-components within each "P"? Do they each get equal weighting? Is the people category more important than the planet? Who decides?

Another option would do away with measuring sustainability using dollars or using an index. If the users of the TBL had the stomach for it, each sustainability measure would stand alone. "Acres of wetlands" would be a measure, for example, and progress would be gauged based on wetland creation, destruction or status quo over time. The downside to this approach is the proliferation of metrics that may be pertinent to measuring sustainability. The TBL user may get metric fatigue.

Having discussed the difficulties with calculating the TBL, we turn our attention to potential metrics for inclusion in a TBL calculation. Following that, we will discuss how businesses and other entities have applied the TBL framework.

The measures that goes into the index.

There is no universal standard method for calculating the TBL. Neither is there a universally accepted standard for the measures that comprise each of the three TBL categories. This can be viewed as a strength because it allows a user to adapt the general framework to the needs of different entities (businesses or nonprofits), different projects or policies (infrastructure investment or educational programs), or different geographic boundaries (a city, region or country).

Both a business and local government agency may gauge environmental sustainability in the same terms, say reducing the amount of solid waste that goes into landfills, but a local mass transit might measure success in terms of passenger miles, while a for-profit bus company would measure success in terms of earnings per share. The TBL can accommodate these differences.

Additionally, the TBL is able to be case (or project) specific or allow a broad scope measuring impacts across large geographic boundaries or a narrow geographic scope like a small town. A case (or project) specific TBL would measure the effects of a particular project in a specific location, such as a community building a park. The TBL can also apply to infrastructure projects at the state level or energy policy at the national level.

The level of the entity, type of project and the geographic scope will drive many of the decisions about what measures to include. That said, the set of measures will ultimately be determined by stakeholders and subject matter experts and the ability to collect the necessary data. While there is significant literature on the appropriate measures to use for sustainability at the state or national levels, in the end, data availability will drive the TBL calculations. Many of the traditional sustainability measures, measures vetted through academic discussion, are presented below.

Economic Measures

Economic variables ought to be variables that deal with the bottom line and the flow of money. It could look at income or expenditures, taxes, business climate factors, employment, and business diversity factors. Specific examples include.
1. Personal income.
2. Cost of underemployment.
3. Establishment churn.
4. Establishment sizes.
5. Job growth.
6. Employment distribution by sector.
7. Percentage of firms in each sector.
8. Revenue by sector contributing to gross state product.

Environmental Measures

Environmental variables should represent measurements of natural resources and reflect potential influences to its viability. It could incorporate air and water quality, energy consumption, natural resources, solid and toxic waste, and land use/land cover. Ideally, having long-range trends available for each of the environmental variables would help organizations identify the impacts a project or policy would have on the area. Specific examples include.
1. Sulphur dioxide concentration.
2. Concentration of nitrogen oxides.
3. Selected priority pollutants.
4. Excessive nutrients.
5. Electricity consumption.

6. Fossil fuel consumption.
7. Solid waste management.
8. Hazardous waste management.
9. Change in land use/land cover.

Social Measures

Social variables refer to social dimensions of a community or region and could include measurements of education, equity and access to social resources, health and well-being, quality of life, and social capital. The examples listed below are a small snippet of potential variables.
1. Unemployment rate.
2. Female labour force participation rate.
3. Median household income.
4. Relative poverty.
5. Percentage of population with a post-secondary degree or certificate.
6. Average commute time.
7. Violent crimes per capita.
8. Health-adjusted life expectancy.

Data for many of these measures are collected at the state and national levels, but are also available at the local or community level. Many are appropriate for a community to use when constructing a TBL; however, as the geographic scope and the nature of the project narrow, the set of appropriate measures can change. For local or community-based projects, the TBL measures of success are best determined locally.

There are several similar approaches to secure stakeholder participation and input in designing the TBL framework; developing a decision matrix to incorporate public preferences into project planning and decision-making, using a "narrative format" to solicit shareholder participation and comprehensive project evaluation, and having stakeholders rank and weigh components of a sustainability framework according to community priorities. For example, a community may consider an important measure of success for an entrepreneurial development program to be the number of woman-owned companies formed over a five-year time period. Ultimately, it will be the organization's responsibility to produce a final set of measures applicable to the task at hand.

Variations of the Triple Bottom Line Measurement.

The application of the TBL by businesses, nonprofits and governments are motivated by the principles of economic, environmental and social sustainability, but differ with regard to the way they measure the three categories of outcomes. Proponents who have developed and applied sustainability assessment frameworks like the TBL encountered many challenges, chief among them, how to make an index that is both comprehensive and meaningful and how to identify suitable data for the variables that compose the index.

The Genuine Progress Indicator (GPI), for example, consists of 25 variables that encompass economic, social and environmental factors. Those variables are converted into monetary units and summed into a single, dollar-denominated measure. Minnesota developed its own progress indicator comprised of 42 variables that focused on the goals of a healthy economy and gauged progress in achieving these goals.

There is a large body of literature on integrated assessment and sustainability measures that grew out of the disciplines that measure environmental impact. These are not constrained by strict economic theory for measuring changes in social welfare. Researchers in environmental policy argue that the three categories; economic, social and environmental need to be integrated in order to see the complete picture of the consequences that a regulation, policy or economic development project may have and to assess policy options and tradeoffs.

Who Uses the Triple Bottom Line?

Businesses, nonprofits and government entities alike can all use the TBL.

Businesses

The TBL and its core value of sustainability have become compelling in the business world due to accumulating anecdotal evidence of greater long-term profitability. For example, reducing waste from packaging can also reduce costs. Among the firms that have been exemplars of these approaches are General Electric, Unilever, Proctor and Gamble, 3M and Cascade Engineering. Although these companies do not have an index-based TBL, one can see how they measure sustainability using the TBL concept. Cascade Engineering, for example, a private firm that does not need to file the detailed financial paperwork of public companies, has identified the following variables for their TBL scorecard.

1. **Economic**
   Amount of taxes paid.

2. **Social**
   Average hours of training/employee.
   From welfare to career retention.
   Charitable contributions.

3. **Environmental/Safety**
   Safety incident rate.
   Lost/restricted workday rate.
   Sales dollars per kilowatt hours.
   Greenhouse gas emissions.
   Use of post-consumer and industrial recycled material.
   Water consumption.
   Amount of waste to landfill.

Nonprofits

Many nonprofit organizations have adopted the TBL and some have partnered with private firms to address broad sustainability issues that affect mutual stakeholders. Companies recognize that aligning with nonprofit organizations makes good business sense, particularly those nonprofits with goals of economic prosperity, social well-being and environmental protection.

The Ford Foundation has funded studies that used variations of the TBL to measure the effects of programs to increase wealth in dozens of rural regions across the United States. Another example is RSF Social Finance, a nonprofit organization that uniquely focuses on how their investments improve all three categories of the TBL. While RSF takes an original approach to the TBL concept, one can see how the TBL can be tailored to nearly any organization. Their approach includes the following.

1. Food and Agriculture (economic): Explore new economic models that support sustainable food and agriculture while raising public awareness of the value of organic and biodynamic farming.

2. Ecological Stewardship (environmental): Provide funding to organizations and projects devoted to sustaining, regenerating and preserving the earth's ecosystems, especially integrated, systems-based and culturally relevant approaches.

3. Education and the Arts (social): Fund education and arts projects that are holistic and therapeutic.

Government

State, regional and local governments are increasingly adopting the TBL and analogous sustainability assessment frameworks as decision-making and performance-monitoring tools. Maryland, Minnesota, Vermont, Utah, the San Francisco Bay Area and Northeast Ohio area have conducted analyses using the TBL or a similar sustainability framework.

Policy-makers use these sustainability assessment frameworks to decide which actions they should or should not take to make society more sustainable. Policy-makers want to know the cause and effect relationship between actions projects or policies and whether the results move society toward or away from sustainability. The State of Maryland, for example, uses a blended GPI-TBL framework to compare initiatives for example, investing in clean energy against the baseline of "doing nothing" or against other policy options.

The European Union uses integrated assessment to identify the "likely positive and negative impacts of proposed policy actions, enabling informed political judgments to be made about the proposal and identify trade-offs in achieving competing objectives." The EU guidelines have themselves been the subject of critique and have undergone several rounds of improvement. The process of refining the guidelines shows both the transparency of the process and the EU commitment to integrated assessment.

Regional Economic Development Initiatives

The concept of the triple bottom line can be used regionally by communities to encourage economic development growth in a sustainable manner. This requires an increased level of cooperation among businesses, nonprofit organizations, governments and citizens of the region. The example in United States show ways the TBL concept can be used to grow a region's economic base in a sustainable manner.

In 2009, the mayor of Cleveland convened the Sustainable Cleveland 2019 (SC2019) Summit to bring together hundreds of people interested in applying the principles of sustainability to the design of the local economy. The SC2019 is a 10-year initiative to create a sustainable economy in Cleveland by focusing on a TBL-like concept. The city uses four key areas for measuring sustainability; the personal and social environment, the natural environment, the built environment (e.g., infrastructure and urban growth patterns) and the business environment. Each key area has six goals. At this point, specific measurement indicators have not been fully developed; however, the city is looking to create a dashboard that could be combined to create an index for overall project success. This dashboard would allow for quick year-to-year assessment in the SC2019 progress.

The Triple Bottom Line concept has changed the way businesses, nonprofits and governments measure sustainability and the performance of projects or policies. Beyond the foundation of measuring sustainability on three fronts; people, planet and profits, the flexibility of the TBL allows organizations to apply the concept in a manner suitable to their specific needs. There are challenges to putting the TBL into practice. These challenges include measuring each of the three categories, finding applicable data and calculating a project or policy's contribution to sustainability. These challenges aside, the TBL framework allows organizations to evaluate the ramifications of their decisions from a truly long-run perspective.

Chapter 34: Conclusion

Even so, many companies are convinced that the more environment-friendly they become, the more the effort will erode their competitiveness. They believe it will add to costs and will not deliver immediate financial benefits.

Talk long enough to CEOs, particularly in the United States or Europe, and their concerns will pour out. "Making our operations sustainable and developing "green" products places us at a disadvantage vis-à-vis rivals in developing countries that don't face the same pressures." Suppliers can't provide green inputs or transparency; sustainable manufacturing will demand new equipment and processes; and customers will not pay more for eco-friendly products during a recession. That is why most executives treat the need to become sustainable as a corporate social responsibility, divorced from business objectives.

Not surprisingly, the fight to save the planet has turned into a pitched battle between governments and companies, between companies and consumer activists, and sometimes between consumer activists and governments. It resembles a three-legged race, in which you move forward with the two untied legs but the tied third leg holds you back. One solution, mooted by policy experts and environmental activists, is more and increasingly tougher regulation. They argue that voluntary action is unlikely to be enough. Another group suggests educating and organizing consumers so that they will force businesses to become sustainable. Although both legislation and education are necessary, they may not be able to solve the problem quickly or completely.

Executives behave as though they have to choose between the largely social benefits of developing sustainable products or processes and the financial costs of doing so. But that is simply not true. We have been studying the sustainability initiatives of 30 large corporations for some time. Our research shows that sustainability is a mother lode of organizational and technological innovations that yield both bottom-line and top-line returns. Becoming environment-friendly lowers costs because companies end up reducing the inputs they use. In addition, the process generates additional revenues from better products or enables companies to create new businesses. In fact, because those are the goals of corporate innovation, we find that smart companies now treat sustainability as innovation's new frontier.

Indeed, the quest for sustainability is already starting to transform the competitive landscape, which will force companies to change the way they think about products, technologies, processes, and business models. The key to progress, particularly in times of economic crisis, is innovation. Just as some internet companies survived the bust in 2000 to challenge incumbents, so, too, will sustainable corporations emerge from today's recession to upset the status quo. By treating sustainability as a goal today, early movers will develop competencies that rivals will be hard-pressed to match. That competitive advantage will stand them in good stead, because sustainability will always be an integral part of development.

Seeking an edge in today's competitive business environment? You can boost productivity and profitability by implementing lean sustainable supply chain management strategies. Whether you are looking to improve your company's bottom line or your own career opportunities, participating in some form of lean sustainable supply chain management initiative is an excellent way to develop and expand your skills in this vital area.

No matter how well a business is performing, substantial improvements can be achieved by applying the principles of lean sustainable supply chain management to all of its practices and processes. That is because the essence of lean sustainable supply chain management is the elimination of waste. When you build a sustainable lean supply chain, it ensures the smooth outward flow of products and inward flow of profits.

Keep Improving!!

Printed in Great Britain
by Amazon